OUR
LIVABLE
WORLD

OUR LIVABLE WORLD

Creating *the* Clean Earth *of* Tomorrow

MARC SCHAUS

DIVERSION
BOOKS

For more information, email info@diversionbooks.com

Diversion Books
A division of Diversion Publishing Corp.
www.diversionbooks.com

First Diversion Books edition, October 2020
Paperback ISBN: 978-1-63576-720-9
eBook ISBN: 978-1-63576-721-6

Printed in The United States of America

1 3 5 7 9 10 8 6 4 2

Library of Congress cataloging-in-publication data is available on file.

To my parents,
for all their love and support in literally everything I do.

CONTENTS

COVID-19 AND THE ENVIRONMENT

∎

Shortly after the manuscript of *Our Livable World* was completed, tragic news broke from the city of Wuhan in China that a highly contagious, SARS-like virus had migrated from another animal to humans—sending medical teams scrambling and ultimately leading to the outbreak of what we now call COVID-19 (also known colloquially, for the moment, as the coronavirus). Through a combination of early-stage government mismanagement and the sheer power of a virus that spreads quickly and easily, the coronavirus soon escaped the city of Wuhan and spread to neighboring cities, provinces, and eventually surrounding countries in all directions.

As many of us are now beginning to realize, our world in the twenty-first century has become far more interconnected than we ever thought possible. We have become more and more globalized with each passing year. The challenges of one nation—particularly if that nation is a major international player—become challenges for the world at large. Such has been the case with COVID-19, an intercontinental menace that quickly moved from regional outbreak to global pandemic. The new coronavirus shot through European nations like a wildfire, igniting explosive epidemics throughout the Middle East, the rest of Asia, and eventually both Americas.

As I write these words, the virus has now reached essentially every-where on Earth.

Italy was the first country among us to issue a nationwide lockdown, sheltering every citizen in place to avoid unnecessary contact and spare their hospital wards an overflow of patients. But the emergency declara-tion was made not nearly soon enough. Intensive care units soon spilled outside of hospital walls into mass-produced tents in nearby parking lots. France would later issue similar orders, requiring anyone venturing out-side of their home to carry paperwork clearing them to be out in public. Neighboring Spain deployed their military personnel to ensure that any-one found outside was either on their way to or from an approved location for supplies or treatment. Spanish hotels became annexed hospitals under the never-ending flow of new patients, while public transit buses became bulk-transport ambulances to fill them.

Iran was especially devastated. Mass graves the Iranian government would later dig to bury their dead were so large they could be seen from space.

Lockdown is currently looking very different around the globe. In Italy, residents with a balcony will sometimes serenade each other in the eve-ning to maintain what *joie de vivre* they can. Apartment tenants have been spotted playing tennis from balcony to balcony, while fitness instructors lead workout classes from adjacent rooftops as student look on from their windows or balconies. The famous canals of Venice, typically crowded with boats and saturated with pollutants, became so clear without daily boat traffic that fish became visible for the first time in years.

In America, the coronavirus spread quickly through highly populated states like New York and California, causing lockdowns and shelter-in-place orders that other states would soon emulate to avoid the worst.

In Canada, experts believe that we are only now—at the time of this writing—beginning to climb the medical response curve that everyone keeps telling us we need to flatten. It would seem that we're slightly out in front of the virus in terms of saving our ICUs from being overrun, for the moment. But folks are already getting into fistfights over toilet paper at Costco, and what many media outlets are now calling "the shut-in economy" is very much in full swing.

The trouble with writing a book about "the good news of technology and our hope against climate change" is that optimism can come off as tone-deaf when offered in a period of deep struggle, sickness, and death.

And we all know that climate change has already carried a heavy human cost. So I have attempted not to be tone-deaf in this work, even if incredible academic teams and trail-blazing tech companies genuinely do have great new options to help with our climate woes. But a global scourge like the new coronavirus ought to remind us that the luxuries we enjoy with twenty-first century technology can also empower a host of downsides that we need to navigate responsibly.

Claiming that there are upsides to a global pandemic can seem equally tone-deaf. The death toll from the coronavirus has already been significant—and it is still rising as I write. I have not yet personally suffered the death of a loved one due to complications from COVID-19, so striking a connection with anyone who has suffered will be difficult. I cannot claim to feel that pain, nor any sense of such a profound loss. But in assessing the impact of this virus from a carbon emissions perspective and from a view of new tech developments over the years, I'd like to offer some glimmers of hope to consider if you are still reading this book in a world beset by mandatory social distancing.

From a climate stand-point, the slowing and stoppage of industrial activity following lockdown orders has caused carbon emissions to plummet around the world—and even more dramatically as fewer folks are driving to their jobs or going away on weekends. In fact, there are practically no cars on the road at all in the areas still strictly sheltering in place. Fewer people are using energy-intensive resources in our cities, and several carbon-heavy industries have all but ground to a halt. Which, in terms of reducing our emissions, has been a temporary reprieve for the environment.

Satellite imagery has allowed scientists to verify that lockdowns and quarantines have had this pronounced effect on reducing emissions. And again: sickness and recession-induced carbon reductions are nothing that those of us who care about the planet ever hope for. But between the time that China locked down their affected cities and right now in mid-April 2020 as I write this piece, China's emissions dropped by a full quarter.[1] Coal consumption at power plants was down by 36 percent without China's massive industries in operation. That is really quite significant: China contributes roughly 30 percent of our global carbon emissions every year. A drop that stark is akin to more than half the entire *annual* emissions from the UK alone.[2]

Every virus-induced stoppage in industrial activity, electricity production, or transportation has resulted in a net loss of carbon emissions world-wide. Commercial air travel, for example—which contributes roughly 2 percent of the world's carbon emissions—plunged as travelers stayed home. Compared to previous years, data provider OAG reported that aviation reductions departing from mainland China dropped between 50 and 90 percent capacity throughout quarantine periods.[3] And when many other nations' borders were closing to prevent incoming infections, those who sought to travel were restricted from doing so. As major events within countries were cancelled to prevent the spread of the virus, the number of flights was reduced even further. Analysts expected that major event cancellations like South By Southwest and large sporting events, involving flights in from all around the country and from multiple industries, saved the world thousands of tonnes of CO_2 emissions from air travel.

As New York state swelled with coronavirus cases and social distancing efforts went into fuller effect, an estimated 10 percent drop in CO_2 emissions over the city occurred.[4] Globally, one estimate from the International Energy Agency (IEA) looking at the numbers in early March suggested that the outbreak would potentially reduce up to half a percent of worldwide oil demand leading through to September.

At the moment, major studies analyzing the emissions impacts of lost industry are hard to come by. But historically, we know that past financial upsets and subsequent emissions reductions have not been very positive in the long run. We typically tend to see a drop in emissions when the markets drop—as in 2008–09 during the global financial crash—though we then see emissions skyrocket during the recovery phase of government stimulus spending. Environmental concerns are often placed on the back-burner when folks understandably just want a return to normal life. Indeed, shortly after the '08–09 financial crisis, carbon emissions rose by about 5 percent as a result of boosted fossil fuel use.[5]

What may be more important, then, is that a dip in global emissions throughout our self-isolation period can simply buy us some crucial time as we develop the technologies discussed throughout this book—especially the technologies with direct implications for advancing coronavirus treatment and vaccine research, like bioengineering and data sciences.

Given the stakes of advancing new COVID-19 treatments and solutions, the global academic community has entered into a new phase of cross-communication. Chinese officials may have bungled their initial response to the virus by withholding crucial information, but the country's scientists would later sequence the COVID-19 genome and release that information online for every lab in the world to use. Since then, an encouraging new trend of doing "emergency science" has proven that we can generate new knowledge very quickly when we need to.

For example, a veritable avalanche of raw data is being released every day by scientists around the world. These daily data sets are made available on "preprint servers" open to everyone, prior to academic publication. Labs never used to have access to such open source knowledge, even just a few years ago during the Ebola scare (when labs were sometimes encouraged to sit on key results until a prestigious publisher would accept them). And looking through these daily data dumps is actually very inspiring. One will find countless authors involved in every new study, with sources spanning universities from nearly as many countries. It is inspiring because, as we now find ourselves facing the same challenges worldwide, there is a rising tide of shared human knowledge beneficial to all.

As new research has streamed onto preprint servers—around ten studies per day, according to one scientist who called the daily deluge "a firehose" of information—teams of scientists would then assess these listings on the fly over platforms like Twitter for mutual vetting and to weed out any pseudoscience. Layman's commentary was then directed to mainstream media platforms for expediency and faster public awareness. Only once or twice, by this author's count, did that result in scientists' unintentional spread of misinformation. Far more often, as a result of multiple academic publications taking on more staff to pump out papers at wartime speed, the tide did indeed lift all boats. Open source viral genomes were posted on academic servers to help labs around the world improve their testing groups (we quickly learned that COVID-19 shares about 80 to 90 percent of its genetic code with the SARS virus that broke out in 2002–03, for example). In fact, according to one scientific commentator writing at *Science*, the level of communication bridged between global scientists has been simply unparalleled.[6]

Academic publishers also tend to prioritize releasing information that will generate headlines and prestige—so, not the studies generating null results. As a result, labs around the world often engage in plenty of redundancy working on challenges that other teams have already failed on but did not publish anything about. Not anymore, thanks to the earlier dissemination of research results. Scientists around the world have been gaining equally crucial insight into what does *not* work with the virus.

With more communication among labs, speedier publication timelines, and increased research-and-development funding for labs worldwide, new science is advancing faster than ever. This has helped, in part, to create the vaccine timeline that experts seem to be predicting for COVID-19. Vaccines typically take two to five years of development before public availability, making the estimates of roughly 12 months for a potential coronavirus vaccine impressive—even if that sounds like an eternity while we're self-isolating and learning of those we love falling ill and dying.

Even more encouraging, a new line of thinking has emerged in some of the economic elite: fund precautionary science now to save ourselves the trillions of dollars lost during a global shutdown. Bill Gates, for example, has funded the opening of multiple factories to churn out crucial vaccine data, even though he knew in advance he would lose billions of dollars on the venture.

Ultimately, our current challenge in dealing with COVID-19 is a lot of things—scary, sobering, life-altering, utterly tragic at times—but our experience in dealing with a global pandemic in 2020 can also offer us some crucial lessons for how our species could deal with other large problems. We've seen the worst in early-stage government mismanagement and selfish members of our communities hoarding the resources we all need. But we've also witnessed how lowering our carbon emissions and reducing atmospheric CO_2 may well look more manageable after the world's populations suffer through this shared challenge. Many of us are seeing the vast impact that individual actions can carry on a global scale. We are seeing the tangible results of governments with shared goals cooperating to manage mutual problems. We are seeing the adaptability of market forces to solve key problems with automakers now working on ventilators and distilleries delivering hand sanitizer. We are learning important lessons about the policy levers at our leaders' disposal for empowering at-risk businesses suffering through difficult change.

We are also learning the kinds of things we can choose to live without. But perhaps most importantly, we are seeing the immense potential in our academic and medical heroes around the world—and the powerful advancements we can all gain from pooling our intellectual resources together. In short, we are witnessing the impact of collectively carving out a future in which we rise to the moment of our largest challenges and finally behave as though we're in this together.

Because we truly are.

OUR
LIVABLE
WORLD

CHAPTER 1

SCENES FROM THE CLIMATE CRISIS

■

I want you to act as you would in a crisis. I want you to act
as if our house is on fire. Because it is.

—GRETA THUNBERG, DAVOS (2019)

I n mid-2019, an ominous headline suggesting that our final few decades were upon us as a species circulated on several popular news media sites.

HIGH LIKELIHOOD OF HUMAN CIVILIZATION
COMING TO END BY 2050,
REPORT FINDS.[1]

The report, produced by an Australian think tank, described in unsettling detail what we might expect in our near future if world governments do nothing about climate change. As you can imagine, the scenario becomes downright apocalyptic by mid-century.

It is true that our governments are not doing nearly enough to address climate change. Studies from academic organizations around the world indicate that our current goals—projected greenhouse gas reductions, speed in phasing out fossil fuels, and more—will not be sufficient to avoid the worst repercussions of climate change. Even the attainment of those inadequate goals has not been progressing efficiently. But, while being brutally honest about the challenge we face is necessary and appropriate, it is equally important to ensure that "doomsday certainty" is not the only narrative driving our decisions.

As temperatures rise, the report continued, reflective sea ice up north will melt. Widespread loss of our permafrost will occur and large-scale Amazon drought and dieback will follow. Sea levels will rise rapidly. Cities close to every coast will be inundated with flood waters. Multiple World Heritage sites will be lost forever. Residents who live in coastal communities will essentially become climate refugees, forced to flee from their homes to neighboring cities and towns—possibly to neighboring countries, if enough of their homeland becomes uninhabitable. Global coalitions will likely fracture and dissipate in the face of resource scarcity and humanitarian crises. Biodiversity will collapse and subsequently implode as mass extinctions occur worldwide.

It's important to note that while these premonitions are unlikely to pass in full, they are certainly not impossible. If our governments truly were to do nothing about climate change—or to continue our woefully subpar response from the last decade—some or all of these prophecies could very well be fulfilled. But many of the most dire, apocalyptic scenarios sound more probable than they should, undoubtedly due to the drama-driven news cycle our major media networks tend to use. Amidst all the countless social media shares, viral headlines, and alarming reports—where most of us get our climate change information—plenty of hopeful news gets lost in the shuffle. Thoughts about climate change then seem to shift from a narrative of possibilities and innovations to the never-ending cycle of chaos, leading us inexorably to the end of our civilization in just a few decades.

Many of us in North America may picture the California wildfires and the rolling electrical blackouts. Some of those blackouts, we'll remember, were reportedly triggered in pre-emptive attempts to spare ourselves from power-line sparks igniting dry forests. Or perhaps we might recall the scenes of floodwaters devouring levees across America following extreme rainfalls and rising tides, filling not just coastal cities but low-lying towns many miles from the coast. Then there are the images of flood survivors canoeing or kayaking to and from their homes in New Orleans through suburban neighborhoods—or possibly stranded on their residential rooftops with spray-painted pleas for help splayed across bed sheets, calling out to rescue helicopters passing overhead.

Or perhaps we picture throngs of protesters hitting the streets demanding that more be done to address our crisis. Recall the Extinction Rebellion protesters staging massive "die-ins" and mock funeral processions for the victims of future climate disasters. Maybe we envision climate protest signs appearing closer and closer to home, emblazoned with phrases more prophetic by the day. "Water is the immigrant they should fear," read one sign in New York City, going basically Instagram platinum at the time.

On just one day in September of 2019, there were 2,500 protests scheduled in more than 163 countries on all seven continents. Roughly 4 million people stood in a collective call for action to address climate change. Protestors far too young to be worried about the end of the world held signs equally honest and devastating. "There is no planet B" read one. "You'll die of old age, we'll die of climate change" read another. "The climate is changing, why aren't we?"

Youth climate strikes have ballooned out from teen titan Greta Thunberg's humble beginnings outside the Swedish parliament. At the end of 2018, tens of thousands of schoolchildren around Europe followed her lead, skipping school every Friday to protest climate inaction. By the end of 2019, protests had spread with hundreds of thousands marching in New York alone, over a hundred thousand in London, and nearly 1.5 million across Germany in one day.[2] Streets filled with youth who feel, as Thunberg so perfectly encapsulates, the frustration and sense of despair in watching their world slowly reaching a point of irrevocable damage without even being of age to cast a vote in stopping it.

Or maybe we picture the world's timeless landmarks already damaged by rising tides and swelling rivers. Think of the ancient cathedrals of Venice partially submerged in November of 2019, with the iconic St. Mark's Basilica closed to tourists when its adjacent plaza's outdoor tables and chairs were bobbing in floodwaters.

Or how about the raging Australian bushfires ringing in the new decade, where "firenadoes" and bloody red skies became the new normal? The combined rise in heat, dryness, and the subsequent wildfires created an environment so hot that so-called "pyrocumulonimbus" clouds—bringing with them fire-induced thunderstorms—formed as a result, according to CNN weather specialists.[3] Embers carried by strong gusts of wind continually confounded responders with fires forming on multiple fronts, later converging into massive "super-fires." Local news anchors reported live that they could even taste ashes in the air while filming the locals fleeing to escape boats. The scenes went viral immediately: the charred remains of kangaroos caught up in fences as they fled and the koalas so parched by the blazing heat they readily took sips from water bottles held by firefighters and dispatched soldiers.

This is how many of us see the effects of climate change: through the lens of scattered headlines, only loosely connected through their sense of doom and foreboding. The only theme connecting these images—aside from the shock and awe of our world under siege—is the sense that nobody really seems to have a stable plan for what we're going to do here. The advancement of climate science, it seems, has simply been a never-ending stream of more and more bad news, punctuated by the occasional not-so-bad news, in the words of atmospheric scientist Katharine Hayhoe.[4]

This book was written to highlight that many scientists and organizational leaders out there do, in fact, have viable plans for achieving our climate goals and saving our planet. Incredible new advancements in science and technology have been developed that will be game-changing in our fight to stop the rise in global temperatures before it's too late. My goal is to reinforce, in the paraphrased words of Thunberg herself, that we've been failing—*but have not yet failed*.

The problem, widely acknowledged and broadcast, is that we have pumped too many greenhouse gases into the atmosphere. In short: while we can now live lives of unparalleled luxury compared to past generations, we use very dangerous kinds of fuel to make it all happen. The energy we use for feeding, housing, moving, and entertaining ourselves are largely reliant upon "fossil fuels" like coal, oil, and natural gas—which, when burned to create the energy we need to power nearly anything we do, also release harmful emissions of carbon dioxide (CO_2).

What makes our problem so monumental in scope is that burning fossil fuels is now so entrenched within every level of societal function that swapping out these fuels will require massive infrastructural change. And that's not mere hyperbole—it includes everything from the clothes we put on in the morning to where our breakfast cereal comes from; the car we drive to our day job; and the show on Netflix we stream at night. Nearly all of it has run on fossil fuels over the course of many decades. As a result, we've been pumping emissions of CO_2 and other greenhouse gases into the atmosphere for quite a while.

With an excess of greenhouse gases in the atmosphere, our first order of business will be to stop emitting more—after all, when you find yourself in a hole, the first step is to stop digging. Finding ways to feed, house, move, and entertain ourselves without the emissions should be the first thing we do. Moreover, we need to find energy sources that involve no net increase in emissions when either sourced or used. That's a tall order, but incredible new advancements in science and technology are getting us very close. We'll explore them throughout this book.

Still, reports are clear that transitioning toward a cleaner and more renewably powered civilization will not be enough to save us. Significant damage has already been done. Even if each of us were to procure our power from sustainable solar panels, drive electric cars, and give up meat for good, the existing greenhouse gases in our atmosphere will continue to warm the planet for many years to come. The latest report from the Intergovernmental Panel on Climate Change (IPCC), as of early 2020, indicates that we will need to address those emissions even as we transition from fossil fuels.[5] So our second order of business, arguably more difficult than our first, will be to *remove* a large portion of the carbon emissions we've already released. We have developed incredible new methods to accomplish this feat, but the cheapest, most reliable method for removing a greenhouse gas from the atmosphere is clearly not putting it there in the first place.

Globally, our emissions have continued to rise annually. At the end of 2019, our carbon emissions hit another record high of nearly 37 billion metric tonnes (or gigatons; Gt).[6] But the rise had at least been slowing. Some of the world's largest emitters, the United States and the European Union, both managed to cut their carbon dioxide output. Global emissions from coal, the worst of all fossil fuels, continued to decline around the world. So it is certainly possible to be a top industrial power and cut your emissions. Experts believed the continuing rise was potentially due to more output from China specifically, and more developing areas furthering their use of natural gas.

Data from the World Economic Forum's annual Energy Transition Index tells us that Scandinavian countries are leading the world with renewables. Sweden, Switzerland, Norway, Finland, and Denmark took spots 1 through 5 respectively. Iceland was also in the top 10.[7] The UK is also pushing forward—in 2017, carbon emissions actually fell to levels not seen since 1890.[8]

Incredible new advancements in renewable energy technology developed by scientists and engineers around the world are driving this process. And even more innovations are on the way to help speed up our transition. According to an analysis from Bloomberg, it is already cheaper for two-thirds of the global population to get power by building a new wind or solar farm than building a fossil fuel power plant.[9] Not just better for the environment—but cheaper! In fact, solar energy is increasing so rapidly that one analyst with the World Economic Forum recently declared that the world will add nearly 70,000 solar panels every *hour* over the next 5 years.[10] Many of these panels will increasingly feature new, state-of-the-art upgrades in efficiency and performance, forcing us to reconsider what is possible with solar energy.

The chapters ahead will focus on how new science and new technology are ushering in a much-needed boost of speed for clean and sustainable energy. Chapter two will hit you hard with what we're up against in our changing climate—and a view of how much our world has already changed. It is a sobering assessment, but it's absolutely crucial that we are honest about the size and scope of our problem. Doing so will help to showcase how truly transformative the impending new developments in cleantech really are. Chapter three will outline some of the most amazing new renewable energy innovations on the verge of changing the world.

For example, did you know that research teams have developed next-gen solar energy tech that will not only power your home long after the sun goes down—but can also produce clean drinking water for your family at the same time? Did you know that by focusing on microscopic aspects of solar cell tech, previously unattainable levels of efficiency have been obtained? Or that "thermal panels" may be able to provide even more energy coverage than traditional solar panels?

Did you know that engineers are crafting "smart highways" lined with solar strips to not only illuminate our lane lights at night, but also to wirelessly charge electric vehicles as they pass by? Or that electric vehicles we're used to charging at night will soon be two-way generators capable of supplying power to our homes overnight and charging up with renewable energy during the day?

Wind energy is improving, too. It may sound incredible, but some newer wind turbine models will actually be larger than the Empire State Building. Engineering advancements in turbine blade tech could reduce manufacturing costs up to 90 percent, leading to similar reductions in cost for builders. Innovative new designs are integrating smaller, maneuverable wind energy drones—capable of self-piloting with a satellite uplink harnessing the power of aerial swells at higher altitudes, then relaying that energy down to the grid below through a ground tether.

In chapter four we'll cover even more advanced renewables that are often less explored in mainstream literature. Some are so ground-breaking that they will impact industries far outside energy production. For example, did you know that biofuels—once derided as overhyped and non-competitive—have now been genetically overhauled after years of bioengineering to become a carbon-neutral ace-in-the-hole fuel source being developed by Big Oil? Or that scientists are now exploring a new and emerging field called synthetic biology for ways to design biofuel sources never before seen in nature?

Scientists are also developing new applications for an energy source derived from the most abundant element in the known universe: hydrogen. Advanced fuel cell tech and hydrogen refinement have progressed considerably, bringing us newer possibilities to use it more actively. And speaking of new fuel technology: Did you know that scientists have now developed the means to produce "solar fuel," using rooftop refineries trapping only sunlight and ambient air?

We're even developing remarkable new energy techs like smart clothing. So-called "green wearables" may soon be outfitted with yarn-like zinc ion threading to churn out electrical power whenever bent, stretched, or even washed with water. Sidewalks have been designed to generate power with the "footfall" of pedestrian traffic. And scientists are now trying to utilize the immense amounts of "waste heat" being released by our electronics and appliances. Want a new taste of the sci-fi? Try the incredible use of artificial intelligence to align vast fields of mirrors—perfectly in tune with the sun as it moves across the sky—to concentrate energy on a single point, generating heat in excess of 1,000°C for energy and industry alike.

Chapter five is about the nonstop forms of energy we will use to support grid function on our way to zero emissions. Critics of renewable energy often cite the unreliable "intermittency" of sun and wind as the Achilles heel of using these energy sources. What they do not often discuss, however, are the new advancements in not just energy storage and grid functions to smooth out the peaks and valleys of our energy use, but also the new advancements in essentially never-ending energy supplies of hydro, tidal, and wave technology. And let's not forget the new engineering initiatives to tap the world's ever-present geothermal reserves and our budding new capabilities in nuclear power to augment the mix. The energy game is changing. A next-generation combination of "smart grids," microgrids, and 5G connectivity between devices will forever alter the way we use energy.

Chapter six covers how we will still manage some semblance of "life as usual" without our historically heavy carbon footprint. While we do require some systemic changes to our way of life, they need not mean giving up some of the things we love the most. New advancements in materials sciences and biotechnology offer us the chance to replace some of the most energy-intensive, carbon-heavy industrial systems we use with greener alternatives. Did you know, for example, that bioengineers can now brew clothing threads in labs? Or that algae can produce bioplastic entirely without the toxic infrastructure we normally use? Or that we now have microbes capable of *building things*?

Chapter seven explores feeding the post-carbon world. Our ever-growing global population is often considered another Achilles heel of climate action plans, but advanced, new agricultural technology (so-called ag-tech) offers us several compelling new opportunities to feed our populations cleanly and sustainably. Did you know that bioengineers have fortified not only our soils, but the crops growing in them? Crops will soon be more drought-resistant, heat-resistant, and pest-resistant without harsh chemicals. State-of-the-art, soil-free aeroponic farms can also utilize AI to supply just the right amount of water and nutrients needed to grow crops—indoors and therefore pesticide free—without herbicides or excess nitrogen.

Chapter eight will explore some of the most compelling adaptations we've created for making our homes smart and our cities even smarter. For example, we now have access to windows that not only block specific wavelengths of light and turn into shades at the direction of an app, but can also generate energy at the same time. We can block out more solar heat while still letting in visible light to save on wasteful AC. And indoor solar cell tech can actually tap the power of ambient light!

Research teams from all around the world are hard at work figuring out how we can cool the planet even while continuing to fuel our cars, heat our homes, and feed billions of people. However, as noted, carbon neutrality alone will not be enough to save us. The world will keep warming unless we actively remove greenhouse gases in the near future. Chapter nine will focus on our newfound abilities to do so.

Since the Industrial Revolution, our average global temperature has increased about 1°C (1.8°F). Scientists estimate that if we want to hold the rise in global temperatures steady at 1.5°C (2.7°F) above our prein-dustrial levels, we need to begin our return of atmospheric levels of carbon dioxide to about 350 parts per million (ppm). For scope: In May of 2019, we officially reached 415ppm CO_2 in the atmosphere.[11] Who knows what milestone we will have passed by the time you read this.

Given that we have already done so much damage by not transitioning away from fossil fuels quickly enough, scientists at the IPCC estimate that we will inevitably need to rely on so-called "negative emission technologies." These are largely technological options—high-tech and low-tech—to help us *remove* CO_2 from the atmosphere going forward. Some are as simple as planting more trees to absorb more CO_2. Others are high-tech, like renewably powered carbon capture machines and bioengineered, supercharged plants designed to absorb more CO_2. The former option may even be bolstered by new advancements in our ability to "upcycle" captured CO_2 into high-value commodities and fuels—creating new and evolving market forces to incentivize the capture of carbon naturally.

The goal is to reach a point in which advanced renewables can power our civilizations cleanly and sustainably while we selectively draw down greenhouse gases from the atmosphere. That is what will "solve" the climate crisis. At that point, called *drawdown*, we will essentially turn back the clock to a time before we blanketed ourselves in excess carbon emissions. This is why the international nonprofit Project Drawdown exists: to collect, analyze, and disseminate information on how we will actually solve the climate crisis.

Sadly, the very reason we are developing many of these technologies is because significant damage has already been done. We already have climate refugees. Floods have already destroyed homes. Species are already going extinct at record rates around the globe. Vulnerable communities have already been deeply impacted by heatwaves and droughts and extreme weather. Chapter ten is dedicated to the damage control we will need to engage in as we progress toward carbon neutrality and as we draw down CO_2 thereafter.

It is important to note that science and business typically work in tandem. If a particular scientist or a whole lab spearheads something incredible, they will often commercialize and subsequently market the idea to make some money. As such, we'll be exploring labs and new companies fairly evenly throughout this book. I mention these two entities because

I'm not affiliated with (or invested in) any of the companies discussed. They are often merely companies started by scientists capitalizing on new technology.

Many of the technologies we will cover throughout this book once existed purely in the realm of science fiction. A scientist would not have believed even fifteen years ago that we are now turning landfill waste into *jet fuel* or that we might potentially have renewably powered and AI-managed indoor vertical farms saving up to 95 percent of the water used in traditional agriculture. Scientists of ten years ago would not believe that by 2020 we would be mass-producing the process of turning microbes into tiny self-sustaining factories, capable of building and assembling everything from fuel to food to the fabric threads we weave into clothing. For these reasons and more, rumors of our demise have been greatly exaggerated. This is not to dispel the need for urgent, systemic action in "decarbonizing" our energy and our industry—but to spread the narrative that we finally have hope not anchored by wishful thinking or promissory science. Rather, we now have hope grounded in the increasingly realistic prospect that we truly can save ourselves from the climate crisis and craft a much more livable world—for greater portions of the human population—than we ever have before.

CHAPTER 2

LET'S FACE IT: OUR PLANET IS CHANGING

■

Climate believers, climate deniers, deep in our hearts
we think it will happen somewhere else. In some other
place—we don't actually say this, but we may think it—in
a poorer one, say, Puerto Rico or New Orleans or Cape
Town or one of those islands where the sea level is rising.
Or it will happen in some other time, in 2025 or 2040 or
next year. But we are here to tell you, in this postcard from
the former Paradise, that it won't happen next year, or
somewhere else. It will happen right where you live,
and it could happen today. No one will be spared.

—NORA GALLAGHER, PARADISE WILDFIRE SURVIVOR (2018)

Y ou can save time debating climate change deniers by simply discussing the facts we have empirically measured on record. You don't even
need to use the phrase "climate change" or "climate crisis." Because if
we're just going by the facts of what we measure in the environment, some
things are damn near indisputable. Regardless of your thoughts on the
more extreme scenarios many scientists have postulated, 2019 was officially the second hottest year since we started recording temperatures
systematically. And that's a fact.

As noted in chapter one, historical data tells us the Earth has warmed
about 1°C (1.8°F) since the beginning of the Industrial Revolution.[1] But
the process seems to be speeding up, as the majority of this increase has
occurred throughout the past few decades. The UN's World Meteorolog-
ical Organization (WMO) tells us that the last decade was "almost cer-

tainly" the hottest decade on record.[2] The National Oceanic and Atmospheric Administration (NOAA) helped confirm that, telling us that 8 of the 10 warmest years on record occurred between 2010 and 2020. The single hottest year ever recorded was 2016.[3] Regardless of what one thinks about climate change, those are simply facts.

Within 2019 alone, June was the hottest June ever recorded. July was the hottest month ever. September was the hottest September on record.[4] October was the hottest October on record.[5] November was the second-hottest November on record.[6] In December, when the month tied for the hottest December on record, a man in Australia was able to cook a whole roast pork in his car.[7]

In June 2019, schools across Europe were closed due to the record-breaking heat in Germany, Poland, and the Czech Republic. The hottest temperature ever recorded in France reached 45.9°C (115°F). Heatwaves caused such wide-spread wildfires that soldiers in countries around the world were dispatched to fight them. City fountains normally reserved for artistic appreciation were full of bathers soaking up water to escape the heat. Despite the record high temperatures, though, Europeans were comparatively luckier than folks in parts of the world closer to global hotspots. Some regions in Pakistan, for example, regularly topped 50°C (122°F) for months at a time over the summer.

Reports tell us these trends have also continued into 2020. January 2020 was officially the warmest January on record globally (inching just past January of 2016 by 0.03 degrees).[8] February was then the second hottest February on record.[9]

One could fill an entire book with stories of humans adapting to and suffering through our new excessive heat. Social media shares of meat being cooked in cars, Olympic athletes training for 5K races after the 10K was deemed "too dangerous" due to chance of heat stroke, companies around the world suddenly providing air conditioning *outside* as workers huddled around fans on break, etcetera, were all too numerous. There were simply too many dire excessive-heat situations to cover here. What may matter more, however, are the ways in which heat will impact us going forward.

Nighttime heat has actually risen twice as fast as the rate of heat increase during daytimes, for example.[10] Two journalists from *The New York Times* were dispatched to Phoenix to cover the expanded new nightlife of people adapting to those more tolerable moments outside. Outdoor workers were often beginning their duties at dawn and dusk while hikers and runners moved toward pre-dawn workouts. The city of Phoenix was, as they had heard, turning into a city of vampires. Low overnight temperatures have increased roughly 1.4°F since the year 1895, whereas the daytime increase has only been about 0.7 degrees.[11]

A side note on how rising temperatures will impact workers going forward: NOAA also estimated that increased heat and humidity have already reduced the amount of work people can do outdoors by 10 percent globally.[12] They expect the figure could double by 2050, which is particularly bad news for landscape and construction workers, among others.

Worse yet, even as temperatures continue to climb, urbanization is continually rising around the world. Combine the two together and we may have a potentially deadly compounding of the urban heat island effect (UHI) forecasted to increase in the future. As global average temperatures rise and regional temperatures rise, the "experienced" temperatures within urban areas will feel at least a few degrees hotter. One new study from the Yale School of Forestry & Environmental Studies (F&ES), for example, revealed that the average temperature rise in urban areas will increase from about 0.5°C to as many as 3 extra degrees in some areas.[13] That is bad news for the estimated 60 to 70 percent of the entire world's population who will reportedly be living in urban areas by mid-century.

Now, we can't attribute every heatwave we experience to climate change. But a long and increasingly well-established consensus among scientists is that rising global temperatures will make heatwaves more frequent, more drawn out, and more intense.[14] In just a short time, we can expect more locations to experience heatwaves. France, for example, experiences double the heatwaves now as they did 34 years ago, and this number is projected to double again by 2050.[15]

Predictably, widespread increases in regional temperatures will also drive up the use of air conditioning—leading to an increase in electricity demand and the harmful chemicals present in refrigerants. It will potentially create a deadly and mutual reinforcement loop: greater production and use of air conditioners will involve more burning of fossil fuels, which will only make the temperature rise even worse.

One study recently predicted that electricity for cooling demand will increase from roughly 400 terawatt-hours per year (TWh/year) to nearly 1,400 TWh/year by the end of the century.[16] That even accounts for air conditioner technology becoming more efficient over time. Sure, for the investors out there, it will turn a roughly 50 billion dollar per year industry into a 1.5 trillion dollar one. But how does that translate into increased emissions? Research conducted by the Lawrence Berkeley National Laboratory and the International Energy Agency (IEA) estimated that typical AC units will come to account for over 130 Gt of CO_2 emissions between now and 2050.[17] That will be roughly 20 to 40 percent of our remaining "carbon budget" allowed under the Paris Agreement if we want to stay below 2°C.

In a country like India, which will face soaring temperatures in the middle of this century and extend into the second half of it, currently fewer than 10 percent of households have AC.[18] As the Indian middle class expands and wealth increases, that number is expected to rise dramatically.

The IEA has further predicted an impending phenomenon they call the "cold crunch" in their *Future of Cooling* report. According to their estimates, air conditioners and electric fans already account for nearly 20 percent of the total electricity used in buildings around the world today, but this amount is expected to rise dramatically as emerging economies supply more buildings and residences with AC.[19]

With temperatures rising and air conditioning use predicted to skyrocket, one recent study further declared that areas encompassing up to a fifth of the world's population will exceed heatwaves surpassing the "upper limits of human survivability." This will occur by the end of this century without immediate action to address climate change.[20]

As temperatures rise, soil moisture can evaporate and ambient air can become more arid, drying out plants. Dry vegetation coupled with high temperatures can make for easily ignitable materials essentially anywhere the combination is happening. While we can't blame every new wildfire on climate change, we certainly have much to say about the increased likelihood, intensity, and duration of wildfires going forward. Or, as environmental writer Bill McKibben succinctly phrased it: prolonged drought, then a record heat wave, then a spark.[21] Warmer nights will also contribute to more sustained fire spread.

Throughout 2019 and into 2020, wildfires raged in California, Canada, Australia, Alaska, Siberia (!), and other areas. The Australian bushfires—made worse by the Indian Ocean Dipole (a phenomenon similar to El Niño) and the hot, dry spell it can embolden—burned up enough vegetation to create a carbon pulse of lost CO_2 into the atmosphere. Researchers at NASA reported to *The Guardian* that the spike was roughly half of the country's total annual emissions alone.[22]

You might recall that the Amazon's wildfires became an international crisis when about three football fields' worth of Amazonian trees burned up every minute throughout July 2019 alone.[23] Fires burned so pervasively that black smoke billowed nearly a thousand miles away to blot out sunlight in São Paulo on August 19. Of course, several of the Amazon's worst fires were intentionally set to clear land for grazing and cash crop plantations. But the speed and intensity of flames were undoubtedly worsened by hot, dry vegetation around more densely forested areas. Satellite imagery also later confirmed the link between hotter and dryer temperatures with increased fires.[24] "Our house is burning," declared French President Emmanuel Macron, echoing one of Greta Thunberg's signature phrases.

The Californian wildfires raged for so long that "fire seasons" will likely turn into "fire years," according to the Wildland Fire Leadership Council.[25] Survivors from the Santa Rosa fire recounted the horror of heat so intense that aluminum wheels on their cars had melted and dripped down driveways like rivers of mercury.[26]

In the case of Alaskan wildfires, roughly 2.5 million acres of tundra and snow forest were lost. The fires were visible from space. On CNN, one could view vast swathes of smoke blanketing Alaska, Greenland, and Siberia. In Siberia, the amount of land lost to fires topped 6 million acres over the summer of 2019 (roughly the size of Vermont).[27] Atmospheric scientist Santiago Gassó declared via Twitter that a "smoke lid" from the fire had spread over about 4.5 million square kilometers across central northern Asia.[28]

Fires in our planet's northern regions are troubling for a number of reasons. For one, the fires indicate that more northern areas are becoming as susceptible as others to easy ignition conditions. Thomas Smith, an assistant professor in environmental geography at the London School of Economics, reported to *USA Today* that the fires were igniting far enough north to tap into peat soils—creating a far worse condition for fire responders and for the planet in general. Peat soil holds immense amounts of carbon, and it also allows fires to burn *underground*, popping up in areas responders may not be.

We find ourselves living in a world where wildfires are destined to happen more frequently, spread more quickly, burn hotter, longer, and emanate from more unanticipated sources. Yes, hotter and dryer vegetation will ignite more quickly. But in California, the deadly Ranch and Carr fires were caused by random sparks from things like concrete construction and from a car tire rim scraping against asphalt, respectively.[29] How are you possibly going to plan for that?

With more of the Arctic on fire than ever before, Arctic ice is melting far faster than scientists had anticipated. So fast, in fact, that the Canadian Arctic is reportedly melting roughly 70 years ahead of schedule.[30] Sustained ice mass lost in Greenland throughout 2019 was so bad it reached levels not expected until the year 2070.[31] This is particularly bad news, given that the Greenland ice sheet is so large that if it were to melt entirely, sea levels would rise by an incredible 23 feet.[32] And it's no safer at the South Pole, either. New research tells us the melting rate has literally doubled over the past six years.[33]

It appears that oceanographers and glaciologists only have stories of faster-than-predicted melting to tell. Alaskan glaciers are melting potentially up to 100 times faster than previously thought.[34] Himalayan glaciers are melting twice as fast as they were in the year 2000.[35] This means that the world's glaciers have lost more than 25 percent of their ice between the years 2000 and 2016.[36]

A melting Arctic, of course, also thaws the ground below—which is really bad news. A nontrivial number of scientists are now telling us that we're particularly doomed due to the melting of the permafrost there. In October 2019, Russian scientists retrieving data on an Arctic expedition discovered, for the first time, that methane was present and boiling on the surface of shoreline waters, visible to the naked eye.

Normally, large amounts of methane—roughly 84 times more potent a greenhouse gas than CO_2 in the first two decades after its release—are safely trapped in regional permafrost. Scientists have often calculated the effects of methane emissions globally from its main anthropogenic sources: livestock and fertilizer run-off. The methane trapped in permafrost was thought to likely remain there (hence the name "perma" frost). But increases in global temperatures have begun thawing the suddenly not-so-permafrost. One UN report called this warming the waking of a "sleeping giant" and warned that what happens in the Arctic—does not stay in the Arctic.[37]

Even more alarmingly, dozens of researchers from around the world are also studying underground temperatures. They have found that areas even tens of feet below the ground have increased in all permafrost zones on Earth.[38] Methane is therefore now leaking from thawing soils and sea-beds. As more methane is released, more warming will occur; speeding up the process of more methane release. Scientists fear a "positive feed-back loop" could occur in runaway methane increase.

Fortunately, new research from the University of Rochester published during the time this book was in editing stages revealed that melting per-mafrost in the Arctic is "less likely" to release apocalyptic amounts of methane into the atmosphere than previously believed.[39] While that does not entirely absolve us from one of the worst doomsday scenarios, it does provide us with some small respite.

Alarmingly, though, Russian scientists have already reported that the rate of underwater permafrost melting has doubled over the past 30 years.[40] In fact, in some areas of the Russian Arctic, released methane has reached nine times the atmospheric average. This rightfully still scares scientists. A few years ago, the sentiment was captured succinctly by Danish climatologist Jason Box, who opined on Twitter, "If even a small fraction of Arctic sea floor carbon is released to the atmosphere, we're f'd."[41]

As our permafrost becomes less permanent and ice continues to melt faster than expected, all of this contributes to the sea level rise we will see throughout this century.

According to a devastating study released in 2019, previous estimations of sea level rise have actually been too modest. New research tells us that some 150 million people are currently living on land expected to be below high tide by mid-century.[42] For scale, research indicates that by the year 2050—if we do not address climate change and at least slow the rise in temperature—Southern Vietnam could essentially disappear underwater. That means the 20 million people who live on this land, including much of the capital in Ho Chi Minh City, will be displaced. Shanghai's core streets will be submerged. An incredible 10 percent of citizens in Thai-land reside on land projected to be submerged in such a scenario. Large swathes of Mumbai, India, would all be underwater. Alexandria, Egypt—with us since about 330 BCE—will be lost to the tides.[43]

Unsurprisingly, scientists also tell us the heat absorbed by our oceans has reached record levels. Even as we feel an increase in surface-level temperatures here, the ocean actually absorbs much of the planet's in-crease in heat. By some estimates, the ocean has retained over 90 percent of the excess heat being generated on our planet since the year 1955 (giving some scientists the impression that 90 percent of our climate change story is being ignored).[44]

Just as we experience heatwaves on land, heatwaves can occur underwater. And as it turns out, aquatic heatwaves now threaten the well-being of corals, fish, plankton, and other marine life, and—you guessed it—more frequently than previously thought.[45] Researchers believe that warming oceans have caused mass migrations of fish toward the poles. One study published in *Science* found that warming oceans have impacted sustainable catches significantly, leading to an average decline of about 4.1 percent.[46]

Scientists studying coral health have noted that oceanic heatwaves have increased in frequency, duration, and intensity. In fact, underwater heatwaves have been so intense that some colonies of coral around the Great Barrier Reef have simply died instantly.[47] But for other colonies, sustained temperature heights otherwise result in coral degradation and bleaching.

As sea levels continue to rise and a warmer atmosphere increasingly holds (and subsequently releases) more water, worse floods are expected to increase. In fact, so-called "100 year floods" that are supposed to occur only once every century—or that have a one-percent chance of happening every year—are now predicted to occur every 30 years along the southeast Atlantic and Gulf of Mexico shorelines, or perhaps annually in places like New England.[48] An estimated 40 million people in the United States, for example, already live in areas with a 25 percent chance of their home being flooded before they can pay off a 30-year mortgage.[49]

Many local governments now view these floods as inevitable—and rather than bailing out residents who insist on not moving and re-building after every flood, are simply buying out homes and telling people to move. The "flood-and-rebuild" status quo has increasingly faltered and been abandoned. Continual costs of re-building and flood prevention like sandbagging, setting up community shelters, and cleaning up scattered flood debris become less of a financial shackle.

Worse yet are less obvious risks to human life from increasing flood waters. As they encroach upon chemical storage sites, or the synthetic fertilizers embedded in farm soils, or the suite of deadly chemicals we often have under our own sinks—this all washes outward together into a toxic soup of random materials. This effect is similar to what happens when wildfires manage to snare, say, chemical processing facilities or the asbestos in old homes. The resulting airborne ash then contains thousands of foreign chemicals suddenly in breathable space for many miles.

And there's a flip-side to the extreme rainfall caused by a warmer atmosphere: drought-stricken farmlands suffering from lost soil moisture and local aquifers drying up during periods of extreme heat. Water shortages are reportedly impending over the coming years as temperatures climb. One new study revealed that the world will experience more frequent and extreme droughts and aridity over the coming century.[50] Even a country like the UK, world famous for its rain, is predicted to face significant water deficits by 2050.[51]

Moreover, the melting glaciers will lead to less water gleaned from glacier springs. This is a problem for the literal billions of world citizens who rely on these springs. One study published in *Nature* declared that 20 percent of the world's population (roughly 1.9 billion people) may be at risk from a loss thereof.[52] It's especially troubling because the demand for water will likely rise in hotter years ahead. Approximately 78 natural "water towers" in which mountain stream systems generate and store water in glaciers are at risk. The Indus water tower, for example, fed by the Hindu Kush and Himalayan Mountains, sustains people in China, India, Pakistan, and Afghanistan. All will be at risk of reduced water supplies.[53]

When towns and cities see their reservoirs run dry, the results are not pleasant. City fountains shut off for water conservation. Showers only happen every few days. Lineups at rationing stations become hotbeds of heightened frustrations.

One might also consider the increasing acidification of our world's oceans. As we pump CO_2 into the air, a large amount of it is absorbed into the oceans—driving up the pH balance. Carbon dioxide absorbed at the surface level reacts with water to form carbonic acid, which oceans then disperse throughout bodies of water.

Incredibly, scientists have found that marine ecosystems have undergone an average increase of acidity around 30 percent over the last two hundred years.[54] One study from NOAA scientists and other academic partners found that some coastal waters (off the California coast particularly) are acidifying twice as fast as the global average.[55] Changes like these could stand to dramatically alter local marine ecosystems and seafood industries.

When waters acidify, this initiates a cascading effect in which the calcium carbonate making up mussel shells and coral skeletons can then weaken. Increasingly acidified waters also mute low-frequency sounds, causing harm to marine species dependent upon acoustic signals to communicate and survive.[56]

Evolutionary biology has been remarkably adaptive in changing environments. But sometimes, environments change so quickly that not even advantageous mutations can catch up. This results in a loss of life when pre-existing organisms suddenly cannot cope with new environmental pressures. We then see a stark loss of biodiversity, wiping out particularly susceptible organisms and putting anything relying on them at risk.

According to a United Nations report on biodiversity loss, more than one million species of plants and animals are now at risk of extinction worldwide.[57] Climate change created by the burning of fossil fuels, according to the report, has simply made it too hot, wet, or dry for some species to survive. Sadly, nearly half of the world's land mammals and nearly a quarter of birds have already had their habitats damaged significantly by climate change.[58] Scientists at the University of Arizona have further calculated that as many as *one third* of plant and animal species could be gone in 50 years' time.[59]

Animals crucial for vast food chains are also increasingly at risk. Insects, for example, sustain larger animals, pollinate plants, and help to keep plants healthy in other ways. But new research reveals that we are on pace to wipe out 40 percent of insect species in mere decades through a combination of climate change and agrochemical pollutants.[60]

You'd expect that with data points like these, world governments would be hard at work solving these problems. And yet, some of the world's largest governments are stalling or denying outright that anthropogenic climate change is real.

Political inaction has been due, in part, to dedicated misinformation campaigns undertaken by fossil fuel energy giants. We now understand that they paid large sums of money to hire fringe, "black hat" scientists and internet forum board trolls to disagree with an otherwise robust scientific consensus on climate change. For this, companies like ExxonMobil and BP are currently being sued for technically lying to investors about the dangers of climate change (cases are still ongoing as I write this).

These days, fortunately—while we all know that scientific views can be mixed regarding the scale and scope of climate change—few debate with the 97 percent of scientists who agree the change is being driven by human activity. In fact, anthropogenic causes for climate change through the use of fossil fuels has now reached the "gold standard" level of academic certainty, as ordained by the Lawrence Livermore National Laboratory in California.[61]

The United States under Donald Trump is perhaps *the* example of political inaction regarding climate change. It wasn't enough that President Trump had once declared climate change a Chinese hoax intended to stunt American industry as a private citizen; Trump effectively drew climate denialism from internet forum boards and thrust it upon the country's institutional base. Pulling out of the Paris called Paris Agreement was one such move; the 90-and-counting environmental rules and regulations rolled back at this time of writing another[62]—such as the Obama-era regulation on fuel economy poised to keep 6 billion tonnes of CO_2 from the atmosphere.[63]

On the other hand, the numbers show that Paris was already a bastion of global inaction. Between the signing of the accord and December 2019's climate conference in Madrid where actionable plans were to be presented, greenhouse gas emissions had still risen by 4 percent.[64] That is a slow-down of how much they *could* have risen, but gases were still rising. (Madrid, of course, was another failure: approximately zero tangible outcomes were produced or committed to—despite half a million protesters marching in the streets outside.)[65]

Sadly, the drive to force politicians into action can stutter and misstep amidst all the positive reinforcement of climate protests. The data shows that there are large portions of us who unfortunately let the hopelessness of a colossal environmental challenge get the best of us—while another large portion allows the anomalies of extreme weather to become normalized. It's just the way things are, we think.

"Eco anxiety," for example, has been a diagnosable illness from the American Psychological Association since 2017.[66] The diagnosis captures the prolonged dread imposed by a seemingly inevitable loss involving the world's many great species and landmarks for future generations. Psychologists believe the diagnosis would be more widespread, were it more reported.

It should be no surprise, then, that with climate change affecting so many aspects of citizens' everyday lives, it has been called a "threat multiplier" by scientists monitoring dangers from all sides around the world. And not just because climate change could impact the infrastructure of coastal cities, crop yields feeding populations, and more—but rather because of how interconnected and globalized the world now is. A grain shortage halfway across the world could affect everything from bread prices to beer shortages elsewhere.

American military commanders, for example, view climate change as a potentially catastrophic security threat, despite what the Trump administration might say about the situation. Converging environmental threats can all contribute toward de-stabilizing communities in ways that can invite military involvement. Rising temperatures, subsequent heatwaves, and sustained droughts can lead to water shortages and health crises. Increasing superstorms can lead to fracturing of coastal communities. Any of these effects (or a combination thereof) can lead to civil unrest. Less functional nations face the additional danger of turning into failed states under the stress of environmental damages. Add to that the impact of mass migrations from an area especially beset by climate chaos. All of these could increase the likelihood of a country like America being dragged into ugly foreign wars.

International threat expert Michael Klare writes in his latest book *All Hell Breaking Loose: The Pentagon's Perspective on Climate Change* that even collateral effects of climate change may drag the world's elite armies into confrontation. These effects could include arguments over natural resources in a newly melted and accessible Arctic, crucial water resources crossing borders like the Himalayan glaciers short-changing China or India, and more.

In fact, according to one study from Stanford, the influence of climate on warfare will increase the statistical likelihood of armed conflict by an estimated five times.[67] Over the past several years, to cite just one example, Chinese cargo ship crews have routinely traveled nearly 3,000 miles from home to Russia's northern coast through a newly opened passage officials have dubbed the "Polar Silk Road," which countries may fight over in a warming world.

There could also be civil conflicts at home from displacements due to flooding, lack of resources following superstorm recoveries, and more. Research already indicates that coastal flooding due to either extreme weather or rising seas will lead to mass migrations inland in the US.[68] As another recent study from the University of Colorado Boulder predicted, the United States could see tens of thousands of extra violent crimes every year due to climate impacts alone.[69] Even now, Oxfam has estimated in one of its *Forced from Home* briefings that one person is forced out of their home every two seconds due to climate-fueled disasters.[70] And if these evictions force the wealthy to flee elsewhere, UN experts fear we may experience a "climate apartheid" in which the rest of us are forced to simply linger and suffer during the aftermath of environmental damages.

As I said, one could write an entire book wholly dedicated to the world headlines describing record-breaking heatwaves, the ongoing collapse of species worldwide, the floods wiping out world monuments and landmarks, the rainfall accumulating toxins from unanticipated chemical soups, the droughts parching crucial farmland that feeds the world's population, and all the potential global instability in our lifetimes due to climate change. Indeed, those books already exist. What kind of future can we envision with all this data?

One of the bestselling works regarding climate change by William Vollmann—directed toward a citizen reading his book in the future—is aptly summarized by a review in *The Atlantic* as such:

> *[Vollmann] envisions her turning the pages of his climate-change opus within the darkened recesses of an underground cave in which she has sought shelter from the unendurable heat; the plagues, droughts, and floods; the methane fireballs racing across boiling oceans. Because the soil is radioactive, she subsists on insects and recycled urine, and regards with implacable contempt her ancestors, who, as Vollmann tells her, "enjoyed the world we possessed, and deserved the world we left you."*[71]

Well, not if the world's scientists have anything to say about it. For every headline about impending doom, let me now counter with headlines about progress. About fantastic new developments. About incredible new research initiatives by innovators and inventors consumed with a passion to save the world. Now that we have framed the scope of exactly what we're up against, it's time to outline how we're going to fix it.

CHAPTER 3

POWERING OUR CIVILIZATIONS CLEANLY AND SUSTAINABLY

∎

A giant solar spill . . . is often just called a nice day.

—ANONYMOUS

Every year, respected academic journals honor research papers of unique scientific excellence by issuing them awards in a variety of fields. One of the world's most prestigious journals, *Proceedings of the National Academy of Sciences of the United States of America* (PNAS), drew significant fanfare in 2015 by selecting a paper charting a path for the US to transition toward entirely renewable energy sources. In his comprehensive work with an equally comprehensive title—"Low-Cost Solution to the Grid Reliability Problem with 100% Penetration of Intermittent Wind, Water, and Solar for All Purposes"—Stanford University professor Mark Jacobson brazenly described how the United States could ultimately power itself cleanly, sustainably, and even more reliably and cheaply without the use of fossil fuels.[1]

Naturally, Jacobson's work became global front-page material after receiving the rigorous academic vetting PNAS is known for. The work also drew backlash immediately thereafter when critics called into question some of its primary assumptions. Namely, the assumptions that thermal heat energy alone would just magically replace battery systems as a storage medium—and that the trillions of dollars spent following his plan would be self-replenishing on short order. Jacobson has since updated his figures in calculating what it would cost to transition the world toward clean energy, if you are curious. In his view, saving the world with clean energy only carries a sticker price of about 73 trillion USD.[2] Which, being about 85 percent of the entire world's GDP, reasonably sounds scary to many people.

Now, many readers and scientists understand that decarbonizing the world's energy systems will not be easy or immediately cost-effective. While Jacobson's original paper may have been far-fetched on metrics of timing and up-front cost, many of us do realize that moving away from the dirty energy sources propping up our global civilizations will be difficult. We will require plans that sometimes do stretch into seemingly cartoonish amounts of money, while initiating change sooner than we'd like. The fanfare surrounding bold visions like Jacobson's papers and grand legislative ambitions like the "Green New Deal" in America reveals that people want real solutions that seem to match the scale of the problem. And in terms of climate change, we've been told the scale is nearly apocalyptic.

This begs the question: Are there realistic green options available to power entire civilizations with clean, renewable energy? Will using clean energy mean we'll have to give up our personal vehicles or our air conditioners? Does relying on renewable energy mean we'll have to hitch a ride on a zero-emissions catamaran if we want to cross the ocean, as Greta Thunberg did to set an example?

The next few chapters will focus on our first order of business: finding ways to power our civilization—its necessities, conveniences, and even luxuries—without the emissions. Working on reducing the impact of our cars in every way possible, lowering the emissions of flights overseas, and the carbon cost of running AC—among the many other accoutrements making up twenty-first century life.

The race toward a zero-emissions, post-carbon world will therefore begin with finding energy sources that are:

- Obtained cleanly without emissions.
- Stored and transported cleanly without emissions, if applicable.
- Used cleanly without emissions.

The engineering challenge in accomplishing all three is monumental. But we're getting there: scientists and engineers worldwide have developed game-changing new advancements in harvesting renewable energy and storing it for use. New solar energy tech can harness more of the sun's power than ever before. Rooftop panels for your home are capable of storing solar energy for use overnight well after the sun goes down. Wind energy tech can now cost a fraction of the price to produce while generating more energy. Advanced new sciences like nanotechnology, molecular engineering, and synthetic biology are changing our perceptions of the cost to produce goods and use energy—all of which is contributing to an ongoing recalculation process of how feasible a proposal like Jacobson's paper or a Green New Deal might be.

The good news (which is largely what convinced me to write this book) is that we have actually developed technology for non-traditional *sources* of renewable energy. Think genetically engineered algae strains living off sunlight and producing carbon-neutral fuel in the process. Think designer proteins made from scratch for tailor-made organisms producing consumer goods. Think solar reactors producing jet fuel from sunlight and air. Consider floating solar-pad installations out in the ocean harnessing light to drive the seawater conversion of hydrogen for synthetic fuel. Or radiative cooling rooftop panels generating power from the cold of *outer space*. And easily deployable underwater turbines harnessing the essentially perpetual motion of Earth's tidal movements while generating 24/7 grid contributions.

We will discuss all of these technologies and the cumulative power of their contributions toward freeing us from fossil fuels over the course of this book. Taken together, the many new energy techs being developed will make grand zero-emissions plans more feasible.

First, let's cover how new developments are impacting the most energy-intensive and carbon-heavy industries for emissions.

CUTTING EMISSIONS QUICKLY: TARGETING THE WORST OFFENDERS

In terms of our worst areas for carbon emissions, transportation and electricity production easily top the list. According to the American Environmental Protection Agency (EPA), transportation alone constitutes roughly 29 percent of the country's total carbon emissions.[3] That's nearly a third of American emissions dedicated to moving vehicles around. We find this ratio relatively consistent in many other countries also, given that global figures cite transportation being among the highest sources of emissions.[4] And as developing countries continue to develop, more potential drivers will be purchasing cars. Figuring out our transportation issues sooner than later should therefore be among our top priorities.

There are three ways scientists have been working on saving our transport sector: creating carbon-neutral fuels for our current vehicles; designing vast new improvements for next-gen electric vehicles; and rolling out new infrastructure to reduce the impact of our getting around. Carbon-neutral fuels hold an incredible amount of promise, but will also depend on our ability to scale them up to civilizational levels. Electrification for our vehicles is more immediate, so it isn't surprising to see immense amounts of capital and innovation starting there. And the good news is that while electric vehicles (EVs) have been slow to grow, we have again made game-changing new developments with tech upgrades.

New EV engineering feats include nanotechnology vehicle specs for increased lightness and durability; autonomous navigation for self-driving cars; the capability to charge our vehicles wirelessly or even while in motion; renewable energy inputs for increased independence for drivers (plus, the potential to monetize vehicles as "virtual power plants"); so-called "smart roads" that can support our drive in various ways; and more—all of which can potentially reduce emissions by making electric vehicles cheaper, more efficient, and ultimately more attractive for prospective drivers.

Depending on where you live, going electric may still be perceived as elitist or unnecessary. But neither of these stigmas can survive much longer. The more states roll out pro-EV regulations, build the necessary charging infrastructure, and offer incentives for buyers, the more these vehicles become a near-term solution for our transport problems. Many countries and individual regions have EV charging stations along major highways already or have those projects in development. Some of these charging stations even have the promise of being renewably powered. If publicly funded, they could also be free to use. Multiple larger global companies have likewise installed free EV charging stations in their parking lots for employees to use. Or, of course, to lure potential customers to parking lots with the promise of free electrons.

The ability to charge wirelessly opens up surprising new opportunities for EV drivers. As such, scientists have been working on developing the tech behind wireless charging for quite some time. Issues with parking misalignment have been a continual problem, as well as housing components capable of protecting drivers from the associated radiation. But the R&D has soldiered on, with major auto manufacturers now partnering with several wireless charging research teams around the world. And fortunately, we now see some companies rolling out this tech for public use.

In Long Beach, California, wireless EV charging company Wave has already provided specialized charging zones for city buses on public transit routes. The pads are designed to charge buses while passengers shuttle in and out, saving the time of routing to a separate charging facility. Scientists have utilized tech that's similar to how we charge our phones without cords, but with additional features to boost charging signals and correct for imprecise positioning.

A company like Massachusetts' Watertown-based WiTricity (spun out from scientists at MIT) has promised their pads will charge through snow and cement, and even if parking is slightly misaligned. "Park-and-Charge; it's that simple" boasts the WiTricity slogan.[5] The company has partnered with manufacturers to scale up garage-ready charging pads for household EVs possibly even by the time you read this (an example image of WiTricity charging pads operating within a parking garage can also be found in the first photo section, on page 60).

The promise of wireless charging tech for electric vehicles is not simply to save homeowners the hardship of plugging in their car at night—the real promise is encouraging public transit operators to transition with increasing ease. Taxi companies may scale up EV fleet ownership if they know that charging pad locations throughout cities will help them avoid re-routing back to the company HQ for power. Any city's fleet vehicles could also employ this technology. Think, too, of stopping zones by hospitals, schools, and anywhere else people routinely idle. Analyses from experts see the global wireless EV charging market increasing from the $21.8 million it was in 2017 to about $1.4 billion by the year 2025.[6]

Engineers won't just stop at wireless charging, though. What if we could take out more of the charging guess-work by adding wireless charging pads to select roadways and charge our vehicles *as we drive*? Yet another tech that seems to always be "just a few years away," mobile charging has also benefitted from several new advancements.

Engineers in Sweden have trial-tested a slot-car-style rail to be embedded in roads, supplying energy to electric vehicles as they drive overtop. The eRoadArlanda project, now in operation since 2017, outfitted roads with a rail running down the middle of each lane to power vehicles on the go. The rail allows cars and trucks equipped with an extendable charging pad to power up at nearly highway speeds.[7] A contact sensor on the charging arm can detect when the rail below is present, then either lower itself down to draw power or raise itself automatically when the rail is no longer present. Alternating magnetic fields facilitate a power transfer between contactless circuits, utilizing a small band of frequency for an inductive power transfer (IPT). And reportedly, pre-existing EVs without an equipped charging pad could be retrofitted with one.

Sweden's e-road has been a relative success so far, its results tapered only by the limits of where engineers placed charging strips. Fortunately, electrifying patches of road is apparently not as dangerous as it sounds: strip channels are wide enough for water to cascade off and not so wide for motorcycles to be impacted. To hurt yourself personally, you would need to be flat against the road jamming something into the strips (which, fortunately again, nobody has attempted to do yet).

Multiple wireless charging options for roadways are currently being developed in various labs around the world. So-called "dynamic" charging pads will be capable of pad-to-pad energy transfer for moving vehicles. But engineers are also designing "semi-dynamic" charging pads for use while vehicles are temporarily stopped (at red lights and loading zones, for example).[8] Early business models for e-roads have centered on charging drivers by the watt with a vehicle pad's unique identifier synced via app, although commercialization has yet to begin.

International tech giant Qualcomm has invested heavily in developing dynamic electric vehicle charging (DEVC) options. Based on the Qualcomm Halo wireless electric vehicle charging technology (WEVC), the company has developed a DEVC system reportedly capable of charging vehicles at up to 20 kilowatts while traveling at highway speeds (courtesy of their FabricEV charging program by Paris).[9]

All new developments increasing the function and appeal of electric vehicles will help to make them a more sustainable alternative to conventional cars. But realistically speaking, next-gen vehicles—whether fully electric, plug-in hybrid, or simply low carbon gas—can all reduce emissions from our transport sector with advancements in other areas. GPS navigation, computer vision, machine-learning, 5G connectivity, and real-time AI-driven analytics, for example, will all contribute to commercializing vehicles that can be autonomous—another feature that will help decarbonize our transport sector.

You have likely heard about driverless cars, but for most auto market analysts, it seems to be a foregone conclusion that autonomous vehicles

(AVs) will eventually dominate the market. In assessing expert analysis, Project Drawdown estimates that AVs will likely capture a market share of approximately 75 percent of cars on the road by the year 2040.[10]

AVs can contribute to decarbonizing our transport sectors in a number of ways. For example, increased data coupled with connected vehicle systems can cut down on collisions, gridlock conditions, and potentially even the number of drivers on the road (more on that soon). Reducing collisions and idling cars can have direct implications for the footprint of remedying either.

With the help of 5G networks, autonomous vehicles will be able to exchange truly incredible amounts of data with each other and third-party sensors in real time. Just think about it—as you drive, baffling amounts of information will be collected by an AV's sensor systems: where you are from moment to moment, locations for where sensors perceive other cars and objects to be, your speed, your calculated trajectory, the power of speed in braking times, and much more. All of this can be rapidly assessed and analyzed by artificial intelligence— as well as communicated with other units doing the same.

Two cars exchanging the same data points in real time will essentially be able to "see around corners" with the situational awareness of where other cars on the road are. Hence, aside from reducing the number of drivers operating cars under the influence or falling asleep at the wheel, AV applications offer more safety through interconnectivity. We'll also see gridlock improvement when AVs become connected to a larger, smarter set of citywide driving data for route optimization.

Combining a number of futuristic AEV features will result in some fascinating new possibilities for drivers. These options are a little further out—but consider AV applications mixed with wireless charging, for example. Equipped AEVs capable of wireless charging will realistically be able to utilize GPS-navigation capabilities to locate nearby charging stations. Stations sensing impending AVs may then be able to disseminate order IDs for streamlining of service. It seems increasingly likely that we will see a day in which AEVs can be dispatched from home, locate a wireless charging station, charge wirelessly without a driver present, then navigate home. On-board camera systems could foreseeably offer the ability to view it all happening in real time. The benefit, again, is less about increasing laziness and more about encouraging the use of electric vehicles for individuals and fleet owners.

For these reasons, some industry commentators see the possibility of future AV ride-sharing services actually reducing the number of people who even purchase a car. After all, it may one day be cheaper to simply call an AEV from a shuttle service, which may have a fleet in motion at any moment; purchase a ride somewhere for smaller amounts of money than fuel, insurance, and possible monthly car payments; and then send the AV on its way. AVs equipped with charging commands could know in advance whether a pre-set passenger destination will deplete its energy reserves and signal operators that a trip back to a charging station may be necessary first.

Driverless vehicles obviously took a roll-out hit back in 2018 when an Arizona woman was tragically struck and killed by an autonomous Uber. But the tech shows no signs of investment loss, with yearly upgrades pushing these cars closer to commercial deployment. Tesla, for example, has already provided drivers with an autonomous summoning capability for use in parking lots (or anywhere within 200 feet, that is). It might sound spurious for those of us who don't mind walking to our car, but these add-ons will all add to the tech's development over time.

Some cities around the world are already using smaller networks of autonomous fleet shuttle services to get the ball rolling. One such service—driverless and fully electric—operates in Lyon, France, carrying passengers back and forth between set tourist destinations. New York City also features a small AEV network compound within one of the city's tech hubs.

Meanwhile, as we wait for a roll-out of wireless charging options, so-called "extreme fast charging" (XFC) tech may also be approaching sooner than we think to further help make EVs more attractive. Researchers in the US Department of Energy are working on EV batteries capable of achieving a full charge for a 200-mile range in less than 10 minutes.[11] Scientists will reportedly achieve this feat by applying high heats to batteries—something universally advised against in the past. Researchers have found, however, that elevated temperatures with controlled and limited exposure times can actually outweigh the negative impacts associated with side reactions within batteries due to more prolonged heat.[12] But that is a work in progress.

With many of these features primarily benefitting EVs, it is no surprise that major manufacturers are ramping up production. I haven't even discussed so-called vehicle-to-grid (V2G) capabilities yet—a new emerging tech development poised to speed up our transition. But even before discussing that, it is worth noting just how much EV manufacturers foresee the market developing in years to come.

In the year 2010, for example, there were only a handful of EV models on American roads. At the time of this writing in early 2020, there are now over 40. In America, there are already gas stations switching to full electric charging ports.[13]

In the last quarter of 2019 alone, South Korean car giant Hyundai announced it would invest tens of billions of dollars in releasing nearly two dozen EV models by 2025. Major car manufacturer Honda announced it would advance its European EV-rollout plans three years earlier than expected. So, Europeans who want to purchase a new Honda vehicle which runs solely on petrol gas or diesel will have to do so before the year 2022. Volkswagen announced it would also spend tens of billions over the next five years creating an EV version of every vehicle on its roster (with further plans to launch about 70 new EV models total by 2028). Not to be outdone, Volvo announced that by 2025, half of its company's global sales will be electric vehicles.

According to estimates in *Forbes*, a dominant EV market could wipe out demand for nearly 13 million barrels of oil per day.[14] In Europe, Amsterdam has already pledged to outright *ban* all vehicles that are not emission free by 2030.[15] German Chancellor Angela Merkel declared that she wants 1 million charging stations for electric cars across the country by 2030. In Norway, gas- and diesel-powered vehicle sales have been forecast to end by 2025.[16] Ireland has planned to ban sales by 2030.[17]

Numerous major shipping companies have also invested in fleets of electric trucks. And why not? Roughly two-thirds of American trucks travel 20,000 miles or less every year.[18] That is an average of 80 miles per day if driven five days per week and 50 weeks per year—well within the per-day operating range of battery-driven electric trucks on a single charge.

National beverage-purveyor Anheuser-Busch announced a new fleet of EV trucks in mid-2019 with an expected emissions reduction of around 910 metric tonnes of CO_2 (the rough equivalent of taking 200 cars off the road). Other companies soon followed suit. All this was around the time of Tesla releasing its own electric truck, and electric truck newcomer Rivian planning a late 2020 roll-out for its EV truck offering—400 or more miles on a single battery charge.[19] GMC will even be providing drivers with an electric Hummer for those who want to trade gas guzzlers for watt guzzlers.

As we wait for EV technology to transition into carbon-heavy, long-haul shipping operations, however, some countries have experimented with an interim solution: specific lanes outfitted with electric cabling that EV trucks can latch onto. In Germany, at this time of writing, a 10 kilometer stretch of the Autobahn has already been equipped with precisely this capability. Siemens has created a system capable of powering trucks via rooftop attachment at speeds of up to 90km/hour. On-board sensors detect when overhead charging wires are available and utilize them when possible—meaning modular installations of this "eHighway" are possible along select patches of road.

Realistically, we should expect AEV tech to factor heavily into shipping industries of the future: automated trucking companies like TuSimple and Alphabet/Google's Waymo are already far along in mass-producing highway-ready trucks. Cupertino-based Plus.ai already has autonomous tractor trailers hauling several-day trips across the US, including stops for fuel and federally mandated stops for rest for anyone who might be on board or piloting remotely. San Francisco-based Embark has likewise completed autonomous hauling trips cross-country.

EV technology is moving to even more industries. The world's first all-electric ferry, for example, was produced back in 2015 in Norway. The Ampere, as engineers called it, reduced the ferry's emissions by roughly 95 percent (with costs also declining by 80 percent).[20] Another Danish e-ferry, the world's largest, is capable of carrying 30 vehicles and 200 passengers while preventing the release of 2,000 tonnes of CO_2 annually[21]—all of which is great news for decarbonizing our transport sector in a growing number of areas.

Electric e-buses utilized by public transit systems are also already a reality. Yes, the costs will also certainly be significant and delay deployment. After all, you have to develop an improved electric infrastructure to handle the increased load of public transit vehicles using grid power. You have to build the charging stations, in the event wireless charging pads are not available—not only at the depot, but around town. You need to have technicians trained with electric motors rather than internal combustion. You've got to buy the actual fleet of buses. None of that is cheap. In the long term, though, EV buses demonstrably carry a lower operating cost and you bypass the emissions, the noise, and the exhaust.[22]

The public won't need to foot the entire bill, of course. Several US companies see the benefit in bankrolling charging infrastructure with a subsequent payment system for individual users. Some companies financing charge stations take a mobile metering approach with a "cost-per-electron"-style billing platform like Ubitricity. Others, like Volta Charging, go with a free-to-use approach, coupled with digital ads for sponsors who will cover the cost. Other enterprising engineers have also developed mobile EV charging units for service providers who want a dispatchable, no-permits-needed, turnkey solution for adding charging infrastructure anywhere they want.[23]

German automaker Volkswagen has even proposed outfitting underground parkades with autonomous car-charging robot butlers—essentially waist-high, electric-supply boxes on wheels that can shuttle a charging pod to your car even if you're not around, use app synchronization to open your EV's charging port, and insert the cable.[24] The robo-butler (it has little eyes and it's adorable, by the way) then returns to its main station to await beckoning from another customer. Each station will be equipped with several detachable charging pods, deployable as needed. When the charging pod fills your EV battery or you return to your car, the robo-butler will then return to retrieve its equipment and shuttle back to the main station. Presumably, of course, for a small fee. Though if this tech helps EV drivers be confident a charging space will be available, I suspect it will be a commercial success.

A primary caveat for EV critics is that vehicle components are often over-mined in carbon-heavy industries and, being finite materials, potentially not in large enough supply to meet our demands. Moreover, by some estimates, the battery component alone can still generate a substantial amount of waste emissions even before the battery leaves a factory—and that other factors, including where one obtains the energy to charge an EV (whether from fossil fuels or renewable energy sources) can significantly alter an EV sustainability profile. However, we do have positive studies reflecting the recyclability rate of different kinds of batteries. And researchers at the University of Exeter, carrying out a lifecycle analysis of EVs from all around the world, recently concluded that even taking manufacturing, production chains, waste processing, *and* method of charging into account, electric cars are still superior in reducing emissions compared to conventional gas-powered cars.[25]

While EV concerns are certainly valid, there are several hopes for improvement. Two primary reasons will need to wait for the next chapter: advanced materials sciences and molecular engineering have us close to replacing our vast production footprint of typical plastics and metals with much lighter alternatives. Battery progress, especially, will be coming up in the chapter about advancements in energy storage. In the meantime, advancements unique to the transport sector can help. New nanotechnology, for example, shows incredible promise for lowering emissions in ways we might not often think about.

Defining "nanoscience" for those of us unfamiliar with the field would take a lot more space than is available here. The basic idea is that *nano* involves the tiny scale of a nanometer: one billionth of a meter. In terms of nanometers, one human hair is roughly 75,000nm (nanometers) in diameter. Nanotechnology, then, involves manufacturing tech at scales of the very, very, very small.

Scientists have developed specialized equipment to manipulate materials at the nano scale using a chemical scaffolding method to separate molecular structures in predictable ways. As a result, scientists working with nanotechnology have been able to explore the sometimes-surprising new properties of materials scaled down to microscopic levels.

Nanoscience is still a relatively new field, with commercial products containing nanomaterials only having entered markets around the year 1999. Early applications included things like consumer paint products containing nanomaterials that could make car bumpers scratch resistant. Or sunscreens full of nanosized reflector fluids to help us avoid UV light. Newer nanomaterials now include products like couches with nanocoated fabrics that resist wine spills (gotta solve the hardest problems first, right?), scratchproof eyeglasses, graffiti-proof walls with crack-resistant paint, new shirts with nanoparticles of silver that line the underarm portion to kill odor-producing bacteria, and more. In short: manipulating matter at its lowest levels means that we are consistently making better, stronger, more efficient, and previously unthinkable products.

Nanoscience can most improve our vehicles in the production of better batteries and more durable and lighter materials. For batteries specifically, the use of nanomaterials offers benefits for energy and power density, cyclability, and safety.[26] In fact, battery engineering has progressed so rapidly that in 2017, experts at Bloomberg New Energy Finance (BNEF) predicted that reductions in consumer cost would see EVs reach price equivalence by 2026. The next year, they needed to revise that number to 2024 in light of new developments. Another year later in 2019, after even more progress in battery manufacturing, price parity was estimated to happen in 2022.[27]

Lightweight, nano-composite materials can also decrease the weight of vehicle bodies, parts, and batteries, ultimately decreasing the production costs and the fuel or energy consumption necessary to move. Even a 10 percent reduction in weight translates to a 5 percent gain for fuel economy, for example.[28]

Nano-coatings atop body materials can also reduce drag, leading to less lost energy, as well as reduce friction inside moving parts. Swedish company Applied Nano Surfaces cites that a small reduction in friction can help reduce fuel consumption by an additional 2 percent, helping to cut emissions with vehicles gaining more mileage on less fuel.[29] Materials like nano-silicates help improve tire function and lead to even more fuel/energy savings.

Nanomaterials aside, we can decarbonize our transport sector further by equipping vehicles with renewable energy inputs—as in lining the bodies of EV models with sleek, advanced solar photovoltaic (PV) panels.

Of course, "solar-powered cars" may forever be a thing of science fiction in terms of running on solar power alone—but every addition of spare energy is a welcome change for EV drivers. New tech upgrades will continually change this assessment, but the industry is certainly exploring options. In 2017, for example, Tesla CEO Elon Musk derisively told an audience at the National Governors Association that, "the least efficient place to put solar is on the car." These days, Tesla's Cybertruck offers an optional solar panel in place of its truck bed area because, according to Musk via Twitter, it will generate an additional "15 miles per day, possibly more." Moreover, that "adding fold out solar wings would generate 30 to 40 miles per day." Yes, you read that correctly: fold out solar wings.

The Lightyear One is a solar vehicle model more fully integrating PV inputs and potentially hitting the road first, though priced absurdly high at ~$150,000. The car is due out commercially after initial reserve models in 2021. At this time of writing, the more affordable Sono Sion is also not yet on the road—but the company is at least taking pre-orders. Production is slated for late 2020 (though likely now delayed) with a price range more modestly projected at ~$25,000 euro. Audi is also now partnered with

AltaDevices to bring in sleek, flexible solar panels into "panoramic glass roof" applications for new models. Amsterdam-based Squad Mobility offers essentially a solar-powered golf cart designed specifically for short-range urban mobility. Their vehicle is lined with curved solar cell tech, providing urban motorists a way to park in motorcycle/scooter areas and drive almost for free on sunny days.

It is doubtful, at this point, that solar power alone will move your road vehicle very far. But what makes solar paneling on vehicles so attractive is not necessarily moving your car without needing to charge it. The vast, continuously expanding charging infrastructure will help with that. The concept of basically driving a *mobile power plant* is what makes renewably powered vehicles so compelling. You have a self-replenishing source of power while off the grid, and—even more revolutionary—you will essentially have a continuous-charging energy-provider. You will, for all intents and purposes, have a solar panel on wheels able to provide energy to your home parked outside. Or, in times of plenty, energy you can then monetize through a peer-to-peer (P2P) marketplace or simply sell back to the grid at large—two concepts we'll be returning to again in the chapter on energy storage.

The concept of using excess energy in your car battery to help off-set home energy costs can seem counter-intuitive at first. Why? Because most EV drivers (or potential drivers) tend to envision themselves depleting their vehicle battery during the day and then charging it at home during the night. But the ability to charge during the day—say, parked by a charger while at work or lapping up sunlight outside—means having the ability to discharge relatively more battery power at night to supplement home energy. This ability to potentially contribute excess energy from your car to your home or even the grid is called vehicle-to-grid (V2G), which I alluded to earlier.

Crucially, areas with a weaker grid presence may actually require some level of V2G participation. After all, the constant rush of EVs charging on a buckling grid may be enough to overwhelm it. One analysis from 2018 actually found that if just a quarter of a percent of Texas's cars were all EVs and attempted to charge simultaneously, they would torpedo the grid's operations.[30] We will discuss this unlikelihood in an upcoming chapter—but in short, a smarter charging network can simply throttle energy accordingly by utilizing AI to gauge when charging is possible, and to what degree (similar to how our ISPs can throttle our internet connection speeds during heavy traffic times).

▪ ▪ ▪

If we consider that renewable energy sources can sometimes provide more power than a utility grid needs, we might also consider that thousands of EV batteries may at any time be connected to such a grid. Those batteries can assist as part of a grid's built-in energy storage system—which users can benefit from in cheaper energy during times of plenty. And some scientists-turned-entrepreneurs are already experimenting with vehicles impacting the grid. Consider Nuvve, a new San Diego-based company utilizing V2G technology to commercialize stored energy by giving owners the chance to utilize their car as a "virtual power plant."

Nuvve's V2G GIVe (Grid Integrated Vehicle) platform works like this: plug in your car, charge your battery, then make money while it's plugged in and sell the excess energy you don't need. Set a safe cap for how much vehicle energy to reserve for yourself as it sells and later depletes, or just save that excess energy for use at home. Nuvve boasts a high-capacity, bidirectional set of AC and DC chargers able to link vehicles to a cloud-connected app. The app helps drivers calculate how much of a charge they will need or can sell. Moreover, corporate vehicle fleet owners are able to monetize unused cars as the virtual power plants they are, gaining energy renewably and selling that stored energy back when not needed—supposedly, with an ability to control the charging and grid-supplementation of more than ten thousand EVs at once.

Some transport sectors will still be tough to electrify and decarbonize, however. Aviation, for example, will need the most work. The aviation sector has been called one of the world's toughest areas to de-carbonize due to its colossal emissions rate. The industry currently contributes, at this time of writing, roughly 2–3 percent of greenhouse gas emissions worldwide. Andy Newman of *The New York Times* has provided the sobering statistic that 32 square feet of Arctic summer sea ice reportedly melts for every airline passenger who flies 2,500 miles.[31] No wonder the term "flight-shaming" has now appeared in our internet lexicon.

Aviation has actually shown promising progress with new advancements in biofuels. We have fuels far closer to carbon neutral already[32]—and polls of customers even reveal that travelers will also pay more for lighter-carbon fuels, on average.[33] We will discuss aviation biofuels shortly. In the meantime, EV plane engineers have still demonstrated progress in using hybrid planes only partially electrified and in electrifying short-haul flights. For example, the Solar Impulse airplane broke records for achieving 40,000km of flight running on solar power alone in 2016. Of course, it also was not carrying a commercial cabin full of passengers. An Israeli start-up like Eviation, however, has designed new electric aircraft capable of flying up to nine passengers around 650 miles using three electric motors in the tail and one on each wingtip.[34] The plane, *Alice*, is expected to enter service in 2022.

Meanwhile, techs with Rolls-Royce and Siemens have teamed up with Airbus to electrify a BAe 146 aircraft. The craft, which typically carries about 100 passengers, will reportedly have one of its four Honeywell turbofan engines swapped with a propulsion fan powered by a powerful electric motor. They say a partially electric, fully commercial flight will be available before the year 2030.[35]

We also have more new options for decarbonizing our transport sector on the ground. Consider that we are already creating better roads with what can only be described as "smart highways."

Realistically, current highways are just long stretches of open road covered with asphalt. All that open space has potential to be utilized, just as we now line the tops of our buildings with solar panels. Likewise, some roadways can potentially be lined with equipment to generate solar energy and spare urban land for other uses while still generating energy. In doing so, we can power lane lights, digital road signs, or—for any of my fellow Canadians reading this book—we can also use heating strips lining smart roads to melt highway snow during heavy storms (dare to dream, eh?).

While most "solar roads" are being developing in places like China and around Europe, the concept seems to have originated in America. Inventors Scott and Julie Brusaw reportedly first proposed them in 2006 with their SolarRoadways project in Sandpoint, Idaho.[36] Progress on the concept since then has been admittedly slow. Only in 2014 did the world's first solar bike path appear in the Netherlands, sparking other countries and start-up companies to follow suit. Even then, Dutch engineers could only achieve a photoelectric conversion rate of about 8.6 percent, which is lower than that of everyday rooftop panels. France would later introduce the world's first official solar roadway in 2016. The Brusaws would eventually see their ambitions actualized with America's first energy-generating thruway in 2016 just outside of Sandpoint.[37]

Results for smart highways have admittedly been mixed on the high end, disastrous on the low end. The "Wattway" solar road in France has turned into an unmitigated disaster. It generated only a fraction of the energy it promised, given that dirt and leaves continually accumulated on the PV panels. Worse yet: the weight resistance for panels had apparently not been tested for tractors. The results were not pretty. German company Solmove had attempted to build a solar bike path, but issues with placement resulted in both short-circuiting from rainwaters and scorching of the panels from road heat. Both issues, Solmove claims, could have been avoided with alternative designs.[38]

Still, the smart highway concept isn't being abandoned. Companies around the world are experimenting with alternative (read: safer) locations for placing PV cells. Emergency lane strips that see less traffic have been one popular alterative, while highway walls and overpass junctions have been others. Essentially anywhere but where cars typically drive seems to be the pattern for larger success.[39]

Results for other popular solar roads have played out slightly more positively. The Ray, for example, is a smart road in Georgia running a 538 square-foot section of the I-85. The Ray is equipped with PV pavement which powers Georgia's Information Center. Engineers in China, meanwhile, have thus far situated their solar road test sites close to several major cities, adjacent to where additional power sources come together. Chinese techs in Pavenergy and Qilu Transportation, for example, have experimented with smart highway tech and on-route sensor networks for autonomous vehicles in years ahead.[40] Data from solar roads from Qilu Transportation in China report that their roads are able to withstand the heavy traffic of about 45,000 vehicles per day and the panels are still able to generate the energy equivalent of powering both highway lights and roughly 800 homes.[41]

It is crucial that while solar road companies are eager begin experimenting with new prototypes, we all manage our promises and our expectations. You might recall the *Solar Freakin' Roadways!* ad from a few years back in which a man simply yells the words "solar freakin' roadways" while boasting the many accoutrements that electrifying our roads could afford us. Video graphics displayed sensors that detect animals or debris on the roadway ahead. Light-up city park playing zones that can offer multiple different light-based loadouts for sport layouts. And, of course, limitless power from the sun. Want all of that? Then build some solar freakin' roadways, the ad implied. While those concepts went viral, industry experts sensed snake oil. A number of skeptic pieces appeared around the blogosphere, wary of everything from dirt and mud and slush atop panel surfaces to the reality that unmoving and eclipsed solar panels will only capture a smaller amount of solar power.

Ultimately, we've been brainstorming more ways to generate energy in previously unused spaces like highways—in the hopes of cleaning up our transport sector. But we have many more locations that enterprising engineers have considered for solar energy outside of transport. We have sidewalks. We have lesser-used lakes and coastlines on which to place buoyant solar pads. We even have transparent solar cells for building smarter windows. And we'll need to use more of them if we want to put a dent into what the EPA tells us is the second-leading cause of emissions: electricity production, at 28 percent. Good news, then, that we truly are finding more and more places to gather energy.

FROM SMARTER ROADS TO SIDEWALKS AND SURFACES APLENTY

What if I told you that walking around your home could potentially one day help generate power and save you money on your electricity bills? That perhaps, in some future life, we might fashion new technology right into the flooring of our homes to harness the kinetic energy produced when specially designed platforms are compressed by footsteps. Would a slightly lower electric bill help motivate you to get your daily steps in?

Or what if I told you that local communities and large cities might one day line "smart sidewalks" with energy-generating solar strips to power anything from public fountains to road signs. That perhaps, at some point, we might combine that equipment with compression-energy technology to feed the electricity being generated by pedestrian foot traffic into a local microgrid for calculated dispersal to nearby lights and signs?

All of the above are already in development. After all, urban pavement areas have been estimated to constitute roughly 30–45 percent of the average city surface.[42] That's a lot of untapped space. How much energy could possibly be generated by people just walking around? One new company specializing in turning footsteps into voltage—an outfit called Soundpower Corp.—reports that an average walker touring a testing phase smart sidewalk can generate about enough electricity to power 300 to 400 LEDs.[43] While that number in itself is not necessarily game-changing, the cumulative power generated by sustained pedestrian traffic in bustling cities could potentially help power devices and sensors for all kinds of purposes.

London-based PaveGen has more than 200 smart cell pavements located worldwide. Cells use electromagnetic induction with copper coils and magnets triggered by footfall—ultimately to power Bluetooth beacons nearby and charging benches that sync to walkers' phones for analytics. PaveGen's systems can produce up to about 5 watts of power while someone is walking overtop, which is enough to power applications like environmental sensors, LED lighting nearby, and energy able to be allocated to local batteries instead if desired.[44] One PaveGen installation in Birmingham encourages walkers to traffic its pathway as a courtesy to others: the energy powers a USB charging hub at nearby benches. The company is now working with appliance giant Siemens to roll out more smart pavements on a global scale.

Solar sidewalks are also popping up in major cities around the world. Hungary-based Platio, a solar sidewalk company, has produced a model made from recycled materials designed to charge nearby EVs.[45] Platio's team has also developed a theft-resistant model to prevent people from stealing the solar cells.

Surprisingly, trial testing for PV-paneling on sidewalks has also been found to help mitigate some of the heat island effect urban citizens experience.[46] Smart sidewalks will also boast anti-slip textural features, heat resistance (to say nothing of those glorious snow removal capabilities again for those of us in colder climates), and enhanced durability for standing up to cycling on top of generating PV power.[47]

In terms of additional unused space: how about *indoor* solar tech? Swedish and Chinese scientists have recently developed organic solar cells capable of converting ambient indoor light into electricity.[48] Of course, the power produced by these cells is rather low, as much less light is available indoors. But the cumulative energy could also power the many devices our impending internet-of-things (IoT) appliances will utilize. As our electronic network coverage increases with new 5G capabilities and IoT expands to take on potentially millions of connected devices even within short ranges—including sensors of all kinds, ranging from moisture to temperature and everything in between—this will require smaller, cheaper, and independent energy sources. Researchers involved in developing indoor solar cells envision the tech could play a role there. (The internet-of-things, for readers unfamiliar with the term, simply refers to the impending connectivity of millions of household sensors and power sources to the internet and each other—providing us newfound abilities for running all kinds of devices through apps and for devices to assess changes in system activity in real time. For example, having a smart system in one's house keeping track of whether there are occupants in a room to allocate light and heat more efficiently.)

Researchers from Uppsala University in Sweden recently took the indoor solar energy concept even further in developing a new type of dye-sensitized cells. By outfitting indoor PV cells with a special dye, the team was able to harvest a unique spectrum of light emitted by fluorescent lamps and LEDs. The cells developed by researchers were capable of converting up to 34 percent of ambient visible light into electricity, offering us another avenue of powering IoT sensors in smarter offices, homes, greenhouses, and more. The team speculated that in the future, autonomous IoT devices will be self-powered by indoor photovoltaic cells.[49]

Or how about underwater photovoltaics? Research carried out at New York University has revealed that harvesting solar energy underwater is not only possible, but several promising materials exist to potentially do so. The team performing this assessment had pictured the technology helping to renewably power autonomous submersibles traveling close to the water's surface—but literally any application that works will still be a milestone achievement.[50]

Alternatively, scientists have already had great success in equipping select waterways and lakes with floating solar pads on the water's surface. Several countries looking for new solar inputs with little acreage to spare have outfitted bodies of water with buoyant solar pads, able to withstand water movement and the wet environment. Countries with long coastlines have also experimented with buoyant solar tech.

Japan, for example, has built not only the world's first floating solar power plant, but is now home to 73 of the world's 100 largest such plants. For a sense of scale, the largest floating solar plant in Japan now spans roughly 18 hectares and can cumulatively power almost 5,000 homes (along with saving over 8,000 tonnes of CO_2 emissions per year).[51]

Floating tech is a relative newcomer to the solar family. Patents for floating solar pads were only produced back in 2008. And yet, there are now dozens of floating solar installations in various stages of development

worldwide. One floating solar plant in China, another of the world's largest at this time of writing, currently harnesses enough energy to power
more than 20,000 homes.

The Netherlands has also planned a 26,910-square-foot "floating solar
farm" called Project Solar-at-Sea involving half a dozen different Dutch
companies. The Netherlands will have 15 modular "solar islands" on the
Andijk Reservoir in North Holland, cumulatively housing about 73,500
solar panels. They will also feature on-board sensors to track the sun as it
moves across the sky for optimal light exposure.

India, too, has recently announced plans to increase its floating solar
expansion along major coastlines.[52] Another solar array just off the coast
of northern Singapore spans roughly five football fields in length. The
project is estimated to power about 1,250 four-room homes and avoid
about 2,600 tonnes of CO_2 emissions.[53]

As it turns out, there are additional benefits to placing solar panels on
bodies of water besides saving space on land. For one, the surface is cooler
and helps to increase the efficiency of equipment. Some floating tech
advocates cite up to a 16 percent efficiency increase over land-based
equipment.[54] In fact, research indicates that surrounding water—not just
the cool surrounding air—can be actively utilized to cool equipment. Usually, the increase in efficiency can mean extra energy for the panels doing
other things (like potentially using the energy to produce chemicals on-
site—but more on that soon).[55] Floating pads can also reduce the increasing evaporation rate of reservoirs.

Surprisingly, installation of floating tech can actually cost the same or
less than tech being placed on land while enjoying a faster installation. In
some cases, land-based costs like excavation, clearing away other land
features, or doing maintenance around solar structures can be a non-
starter relative to having floating pads nearby.

Floating solar has now gone from being a mere curiosity to generating
gigawatts of energy in large-scale installations over the course of just a few
years. In July 2019, South Korea built the world's then-largest floating
solar farm spanning 30 square kilometers (11.6 miles). They expect the
floating solar farm to generate enough electricity to power about 1 million
households.[56] No surprise, then, that some analysts believe the global
average of floating solar projects will grow by an average of 22 percent
every year from now through 2024.[57] In America, researchers at the American National Renewable Energy Laboratory have estimated that these

"floatovoltaics" could potentially cover up to 10 percent of the country's energy needs if used on just one-fourth of the country's manmade reservoirs.[58]

Of course, volatile weather can make floating solar plants a tricky option to consider anywhere waves can become extreme. One typhoon near a Osaka, Japan plant back in 2017 caused intermittent power outages for systems connected to the network. Placing objects on the surface of lakes and rivers also may block sunlight potentially being used by organisms below. Marine ecosystems may require some or all of the sunlight which would have been shining. As always, research is ongoing to consider these issues and ways we might account for them. But the concept of applying solar technology to unused surfaces, even via waterways if necessary, is one actively being researched and deployed.

NEXT-GEN SOLAR ENERGY TECH

Along with new locations for placements for solar cells, we could certainly extend the solar conversation to general technological advancements. Solar photovoltaic cells have progressed substantially over the years in terms of generating more power, withstanding the elements and being more affordable. But crucially, silicon-based solar panels—and even photovoltaic designs in general—are no longer the only game in town.

Research into improving silicon solar panels has still continued unabated. One innovation from scientists at MIT and the National Renewable Energy Laboratory in early 2020 found that silicon cells could actually be much thinner, saving on substantial manufacturing costs going forward (and subsequent costs for buyers). By advancing the process in which silicon is handled, the team found they could shave down the standard 160 micrometer size of silicon cells to eventually as little as 40 micrometers or less—requiring only about one-fourth as much silicon for manufacturing.[59]

But some of the most exciting new developments in advanced solar PV tech are coming from entirely different materials. Perovskite, a solar-harnessing dust-like material and named after lab-grown crystalline materials that resemble the naturally occurring mineral perovskite, is one such development. Printable just like paper and capable of producing higher solar-harnessing efficiency ratios than traditional silicon, perovskite gives developers promising material to work with.[60] So-called "quantum dot" nanotechnology is another new development. Both applications allow for solar tech to achieve greater efficiency levels in newer lab trials—and especially when combined.

Mainstream silicon panels are durable and reliable, but there is a natural limit to how energy-efficient they can be in converting solar power to electricity. The average solar panel at this time of writing receives about a 15 to 20 percent efficiency rating on average. Perovskite, however, is a material which achieves higher energy conversion rates by virtue of additional "light-harvesting" surface properties. Solar cells with a perovskite coating have been demonstrated in some situations to achieve over 24 percent on average.[61]

Perovskite is still an up-and-coming technology (even though it has been in development for a long time) because it has limitations that scientists are working to overcome. The material is brittle and sometimes fragile, which is problematic when there are potentially polluting materials contained inside. But perovskite is also seeing incredible investments into R&D testing. Leading solar tech provider, UK-based Oxford PV, now provides a "tandem" cell combining both perovskite and silicon to achieve roughly 20 percent greater efficiency than traditional silicon-only panels.[62] Analysts expect that this will become the new industry norm for the time being—dual perovskite and silicon cells.

Danish solar tech start-up InfinityPV, meanwhile, essentially makes solar tape. Materials are produced from a printer-like apparatus onto the non-sticky side of tape which is then sold to consumers. Anyone using the product can then apply the tape to a surface featuring solar-generating material on the side they see, needing only to attach a converter (which InfinityPV also sells) to the end of the roll they have cut. Some applications might involve soldering and even calculating voltage, however, so it's only for the DIYers for now. But ideally, picture similar-minded applications creating solar-coated tarps in a hardware store near you.

Researchers mingling nanotechnology and solar cell tech have found a range of improvements by fine-tuning panel parameters. Nanotech applications for creating hydrophobic nano-coatings that repel water have helped to improve the energy produced by PV panels. Even newer silica nanomaterials produced with anti-reflective and superhydrophobic coatings offer panels a 10 percent improvement in performance, self-cleaning capabilities, and an additional 5 percent increase in absorbed wavelength range.[63]

More recently, scientists at Rice University were able to increase solar efficiency with another nanotech innovation: carbon nanotubes. These structures are essentially nano-sized tubes of honeycomb-shaped carbon atoms, all attached to form a lattice pattern. By integrating this material, the team at Rice was able to channel some of the waste photons (read: sunlight) being reflected off panels as thermal radiation. In doing so, the team was able to raise cell efficiency by about 22 percent.[64]

Scientists investigating applications for greater solar efficiency have also been able to manipulate materials at the nano level to direct greater amounts of sunlight onto usable portions of the cell. This includes quantum dots, which are tiny semiconductor particles scaling at only a few nanometers in diameter applied to PV panels. When these dots are precisely the right size, incoming light waves will cause the electrons of dots to "jiggle" (read: achieve plasmonic resonance) at a frequency that achieves higher-efficiency energy transfer. As PV expert Varun Sivaram has described it: the effect of exciting electrons in this way is rather like an opera singer hitting the exact right note which shatters a champagne flute.

Tailoring quantum dot applications for solar energy will offer engineers some unique new opportunities. In early 2020, for example, scientists at The University of Queensland announced they were able to produce a transparent, flexible solar skin material able to wrap the contours of other structures. Their quantum dot application produced a world record for converting solar energy into electricity at the time, generating a 25 percent improvement in efficiency gained.[65] And according to one of the scientists involved, "this opens up a huge range of potential applications, including the possibility to use it as a transparent skin to power cars, planes, homes and wearable technology."

Previous solar panel designs have suffered from a natural ceiling on the amount of energy able to be harvested from incoming light. They have functioned by the energy in photons reaching the solar cell and exciting electrons, which we then channel for use. But photons reaching a solar cell with insufficient energy will not "excite" an electron across the so-called "bandgap," resulting in lost energy which is dissipated as heat. We have improved upon this limitation by using quantum dot technology— with optical properties able to tune the bandgap and allow for a greater yield of solar energy.

New developments are occurring constantly. One new engineering feat for stand-alone solar panels is to make them "bifacial": able to capture sunlight from both the side facing the sun and the side facing the ground. In a rare instance of science serendipitously falling in line with pop culture, researchers reported in late 2019—right around the release of the latest *Star Wars* movie—that they had tapped into "the dark side" of panels. Scientists were able to generate roughly 15 to 20 percent more electricity than today's "monofacial" solar panels. Market analysts expect that enough bifacial solar panels will have been released by 2030 that they will account for roughly half of the tech's market share worldwide.[66]

Or how about solar windows that will also harness solar energy? We will return to "smart windows" later, in terms of selective wavelength shading for massively increased insulation during winter months, as well as sun-blocking capabilities during summer months. For now, in the case of solar energy, scientists have developed multiple window designs capable of generating energy from sunlight while also just being windows. And, as it turns out, solar windows can also offer superior insulation.[67]

Building-integrated photovoltaics (BIPV) have increasingly factored into newer architectural designs as companies seek to gain "net-zero" building status. Or, at the very least, simply pay less on their energy bills. The visible light that we can normally see ranges from wavelengths of light within the 400nm range to 700nm (running violet through red, respectively). Solar windows, on the other hand, can absorb additional wavelengths of light outside this range within the ~700nm and 1mm infrared (IR) range. Scientists have used this differential to create transparent solar cells (TSCs).

Multiple labs around the world working on TSCs have managed to produce a commercialized product. Researchers from the University of Exeter in the UK created "Solar Squared" modules for easy, scalable integration into building designs. A separate team of international scientists even managed to create flexible, bendable, transparent solar cells for use in curved-window buildings. Several international architectural firms are considering using them.[68]

Another academic spin-off company called Soliculture, from UC Santa Cruz, offers solar windows specifically used in greenhouses. Their wavelength-selective photovoltaic system (WSPV) adds a luminescent magenta dye in its panels to absorb the sun's blue and green wavelengths. The result is a conversion into red light, which researchers found can actually yield a higher photosynthetic efficiency for plants.

Soliculture's LUMO panels are already on the market, with results of their light spectrum tech offering some crops a two times greater weight relative to control crops.[69] And yes, the plants still appear green. Plants need only a small window of light wavelengths to grow; absorbing red and blue wavelengths but reflecting green, which is largely why plants appear green to us. But fear not, you greenhouse operators who do not wish to grow in reddish light: researchers from the University of California found that plants will grow nearly just as well under merely transparent solar windows.[70]

MOVE OVER, PHOTOVOLTAICS: THERMAL ENERGY TECH

Now, this is where solar energy innovation really starts to heat up (sorry). Sunlight actually offers us two other compelling alternatives for harnessing renewable energy: concentrated solar power (CSP) and thermal energy units. Both techs involve harnessing the heat energy contained in sunlight and either using it immediately or storing that energy as heat to be used later.

Concentrated solar power plants often include vast fields of mirrors stationed around a central tower, directing sunlight toward it. As every mirror in the array tracks the sun to reflect its rays, the central tower collects the heat from every solar ray in unison. That thermal energy is then either stored as heat or supplied to an adjacent city for renewable solar energy even while the sun is out of commission.

Crucially, though, thermal energy devices can be smaller, modular, and rooftop-ready—a much newer innovation. Consider, as noted earlier, that today's solar tech involves an incredible amount of energy lost as wasted heat. This is something advanced new PV panels are working to overcome. But in the meantime, researchers in Houston have developed a "hybrid" solar device utilizing new advancements in molecular energy and heat storage to create a multi-layered modular unit—able to both capture solar energy and use it right away, or store its heat for potential energy later on through its medium of layered materials.[71] With this approach, the team can harvest a fuller spectrum of light and direct some for use now while banking the rest.

Incredibly, the Houston team's device offers a harvesting efficiency of about 90 percent during the day and an efficiency of 80 percent at night, producing temperatures even hotter than those of the daytime. This is because at night, the stored energy is being harvested by the molecular storage material instead of the capture material. Unlike the capture material, the storage material essentially converts from a lower energy molecule to a higher energy molecule to generate a higher temperature. I wrote to Hadi Ghasemi, an associate professor at the University of Houston, about the device—and he elaborated that the device could certainly be rooftop capable in the future, offering homeowners some level of after-hours energy.

Ultimately, I suspect that new developments in solar energy tech will transpire so quickly that a few may slip by while this book is being edited and released. In fact, as I was originally writing this section back in June of 2019, work published in *Nature* reported that scientists were able to artificially increase the theoretical limit of silicon's energy conversion efficiency—by increasing the number of free electrons able to be harvested by cells.[72] That was new. Quantum dot tech also raises the bar to more electrons, but normally photons striking panels can only free up one electron, max. Scientists found that by getting high-energy photons to strike silicon, they were able to free up two electrons instead of just one. And while that didn't double the energy conversion rate, researchers estimated that the percentage of conversion could rise by several points above the theoretical maximum for silicon (which was already 29.1 percent) to about 35 percent—all of which means that we might not need to entirely "retrofit" traditional panels with fancy new ones if we can just augment them with upgrades. Additional updates between drafts have necessitated more inclusions, because the research is now happening very, very quickly.

It is worth noting, for the sake of highlighting how governments will count on new science and new technology over time, just how much solar energy tech is now scaling. Countries all around the world are investing in solar farms that stretch for literal miles. In the Datong County of China, for example, passersby on the ground will see miles and miles of solar panels numbering into the thousands. If you happen to be flying overhead, you will see that they have been artfully arranged to showcase two giant pandas (one of whom is waving at you). The array spans an astounding 248 acres of land. And while that's certainly impressive, it is also relatively small for a "solar park" comparative to others around China and the world. Another park in China holds roughly four *million* panels, forming a mass so large that you can see them from space.[73]

Elsewhere, in July 2019, Australia announced plans to build the world's then-largest solar farm—worth $20 billion USD—in the country's Northern Territory. The farm will power vast portions of nearby Singapore via undersea cables rounding Indonesia. Construction for the so-called "Sun Cable" project will begin in 2023 and conclude in 2027.

Indian Prime Minister Narendra Modi has already pledged to install as many solar panels in the country as half that exist in the entire world. Plans now include giant, sprawling solar farms adjacent to cities and the lower-hanging fruit of lining panel technology atop the many rooftops in the country's crowded city centers. India's 2018 total investment in solar power actually outpaced coal, according to a report produced by the International Energy Agency.[74] The report indicated that continued investments into solar at that scale will have India well on-target for hitting its Paris climate commitments by 2022, as the country has already approved dozens of solar and wind projects in years to come.

The Middle East, long known for its oil production, has made significant strides toward a renewable transition. Engineers in Abu Dhabi, for example, opened one of the world's largest solar farms in 2019 (if not the biggest, at this time of writing), clocking in at 1,177 megawatts (MW). The installation took two years and upwards of $850 million to build, but now stands as a beacon of renewable energy hope in the region. Abu Dhabi also made international headlines throughout 2019 for activating its long-awaited sodium-sulfur battery plant, capable of corralling energy from 10 different locations as one. This "virtual battery plant" is able to store about six hours of energy in the event of a grid outage for the city.

Though one could write a very depressing book about all the damage we're doing around the world, one could just as easily write a book about all the incremental steps of progress being undertaken. For example: in 2011, the market for solar energy in Latin America was roughly zero. By 2021, Latin America is now estimated to account for roughly 10 percent of the entire world's solar PV demand.[75] In America, some states are literally now mandating that all new homes *must* be equipped with solar power throughout the 2020s and onward. California was among the first to write this requirement into law. Appraisers estimate that the added cost will up the price tags of new homes roughly $10,000—but with reduced monthly energy expenses paying this off long before one's mortgage comes due. In fact, NPR estimated that homeowners will save approximately $19,000 over the 30-year lifespan of one's mortgage period.[76] Moreover, a recent report from Environment America Research & Policy Center revealed that solar power capacity has more than doubled in 45 of America's 57 largest cities over the last six years. They added that roughly a third of American cities had quadrupled their PV capacity.[77]

With more panels being used worldwide, scientists are now tracking some interesting trends in panel placement. Consider the new trend of "'agrivoltaics," for example, involving the combination of solar arrays with agricultural crops or land restoration projects. One recent study published in *Nature Sustainability* found that native plant restoration projects coupled with solar PV panels parsing the cropland can actually increase panel efficiency considerably—and lead to bonuses for crop growth.[78] The combined shade and microclimate adjustments to water vapor in the air appear to benefit both panels and crops, either in the shade or close to it.

Of course, we do also have to consider the negative impacts a massive transition toward solar energy could imply. Aging solar panels will someday need to be replaced—and they do have hazardous materials embedded within them. As we pursue renewables on a global scale to off-set our use of fossil fuels, we'll need to keep in mind the shelf-life of many PV panels is only around 20 years; 30 max. Recycling old solar panels will involve finding ways to interact with exposure to and then storage of dangerous chemicals. But, as always, the good news is that we already have people working on it: we will return to the issue of pollution from renewables and other forms of pollution in the chapter on damage control.

Let's turn now, after our latest transport and solar power efforts, to another large part of our renewable energy portfolio: new advancements in wind energy.

WIND ENERGY

Wind energy has come a long way. Civilizations throughout much of recorded history have attempted to capture the force of wind and turn it into mechanical power. Ancient boats equipped with sails helped to propel us through the water to discover distant, far-away lands. Wind swells pushing against windmill blades have created the mechanical force necessary to pump water and crush grains. And now—since the discovery that coils of metal moved with magnets can generate electrical energy—modern wind turbines have harnessed the power of wind force to create another form of renewable energy.

Whether we're building modern wind turbines onshore in our fields or offshore close to our coasts, new advancements in wind energy tech have evolved to capture more and more energy from the force of air. It hasn't been an easy sell, though. Wind builders have had a hard time convincing skeptics. Early modern designs did not capture incredible amounts of energy. And the wind does not always blow, of course. Hence using wind energy to help supply our colossal power needs in a 24/7 post-industrial society has not been a capturable outcome. The difference, now, is that wind turbines have gotten much larger, cheaper, and more efficient. Since the 1970s, for example, wind turbines have improved to generate roughly 100 times more power than their previous models even as the cost to produce that electricity has dropped from $500 per megawatt-hour to a mere $50.[79]

A brand new, state-of-the-art wind turbine is absolutely massive. Advanced new models will typically boast a rotor blade wingspan of about two *football fields*. This is a remarkable feat of science and engineering. It also makes wind energy less able to be passed off as a mere novelty. Wind energy tech has further improved to tolerate colder climates and, crucially, placement in offshore installations. While offshore installations may be costlier to produce than those on land, they come with a number of hard-to-ignore bonuses. For one, they bypass the pervasive "not in my back yard" (NIMBY) effect that wind farms can have. (President Trump erroneously telling everyone the noise can give you cancer also doesn't help, of course.) But in addition to diffusing the NIMBY effect, offshore farms can also generate more power at steadier rates, with stronger and more predictable wind patterns.

Consider that America's first wind farm was constructed only as recently as 2016 off the Rhode Island coast. At this time of writing, there are now more than 58,000 offshore wind turbines in operation and the number is growing quickly.[80] Coastline real estate for wind turbines is so lucrative that companies like Shell will put down top-dollar bids to acquire "sea bottom leases," just to lease the areas to place offshore turbines.

By some estimates, offshore wind is expected to grow by 16 percent annually through 2020 as installations gather in Canada, the UK, Germany, and the Netherlands. Another new massive offshore wind farm is under construction in Massachusetts, set for completion in 2021. The farm will reportedly save customers about $1.4 billion in electricity savings over 20 years—as well as reduce emissions by an estimated 1.6 million tonnes of carbon dioxide (the equivalent of taking roughly 325,000 cars off the road).[81] Massachusetts further declared a goal of setting up enough offshore wind installations to power roughly 1 million homes by 2027, which legislators then later doubled.

Meanwhile, again at this time of writing, the UK leads the world in offshore wind power capacity with an incredible 34 percent of all total installations.[82] In fact, the world's largest wind farm is currently being built off the east coast of England. Hornsea 1 will feature an array of more than 170 turbines, the cumulative energy output of which will reportedly be capable of powering roughly 1 million homes. Germany has installed 136 new offshore wind turbines around the North and Baltic Seas. They currently account for roughly 28 percent of total installations around the world.[83] Denmark is making the most of its presence in the region: one massive offshore wind farm under construction off the island of Møn in the Baltic Sea will generate enough electricity to power 600,000 households when it is completed in 2022.[84] Energy powerhouse China has also invested in wind: the International Energy Agency has reported that China will soon eclipse others as the world's offshore wind leader.[85] The Global Wind Energy Council further predicted that 2023 world leaders in offshore wind generation will be the UK, Germany, and China.[86]

Onshore wind energy also has incredible potential. One analysis from the University of Sussex and Aarhus University conducted in late 2019 found that there are enough suitable onshore wind installation sites in Europe to power the entire world—year after year—until 2050.[87] And in terms of American wind capabilities, it's worth noting that researchers have analyzed how many new sites devoted to turbines would be necessary to reach America's 2030 wind energy goals. The answer? None. Simply retrofitting and increasing the "installed capacity" at pre-existing turbine sites would roughly quadruple the energy output from 2014 levels.[88] By their calculations, if these retrofits were undertaken for America's existing turbines, the country would already have reached its 2030 goal (which, for reference, is generating 20 percent of the country's electricity). If America reaches that goal, approximately 825 million metric tonnes of carbon dioxide emissions in the electrical energy sector will have been eliminated.[89]

Now, onto developments. The world's largest offshore wind turbine, at this time of writing, first had to be installed on land just for testing due to its size. General Electric's Haliade-X 12 megawatt (MW) prototype installed in the Dutch city of Rotterdam near the end of 2019 stands roughly 853 feet tall with a rotor blade diameter of nearly 720 feet. The prototype immediately set records by generating roughly 260 MWh of clean energy in 24 hours, enough to provide power for 30,000 local households.[90] Commercialization is set for 2021.

Of course, we do reach a cost-benefit tipping point with creating larger and larger turbines. Large components must be shuttled to the site during construction, necessitating roadways. A top-heavy turbine also means an extra-wide (and extra-sturdy) base to support a larger tower and longer blades. As a result, turbines must generate more power over time to offset both the carbon cost going into increased construction and the monetary cost of the same. New innovations in manufacturing could help to make up for it, however.

Many of the newest materials advancements in solar cell tech have also opened up new possibilities in wind. For example, consider that turbine blades can always become lighter but stronger with nanomaterials—and what their increasing use and falling costs can mean for the tech. Engineers at MIT have recently developed a method for producing aerospace-grade composites without all the enormous ovens and pressure vessels, vastly reducing the energy-scale of nanocomposite material going into lighter but stronger blades. Researchers essentially placed materials inside a smaller oven merely to cure them, then wrapped them in an ultrathin film of carbon nanotubes. Applying an electrical current to the film caused the nanotubes to respond like a nanoscale electric blanket, fusing the materials inside together using only 1 percent of the energy.[91]

Innovations in wind tech have been surprisingly diverse. Consider "bladeless" wind turbines that use the force of wind to drive a piston up and down for electricity. Spanish start-up Vortex Bladeless has engineered a whole new class of bladeless turbines. Instead of a turbine's signature blades spinning around, Vortex harnesses the physical power of "vorticity"; the spinning motion of air (and liquids). When the wind hits a bladeless turbine, it passes by several cylinders along the pole, creating a spinning vortex motion. Oscillating cylinders then translate vibration and kinetic energy into electricity. While the energy produced is less than that of a turbine with blades, the lack of gears and bearings makes the bladeless turbine potentially cheaper and able to generate power for less cost. The reduced cost of manufacturing and maintenance may make bladeless turbines an attractive option for ships and rooftops.

Other new wind energy techs have also capitalized on the larger wind-energy ratios found at higher altitudes. A variety of intriguing new designs from energy companies around the world have innovated on capturing high altitude wind energy (HAWE—though some simply prefer AWE for airborne wind energy at various heights).

Consider Project Makani, a spin-off (and now defunct) team that originated in Alphabet/Google's more secretive "X" division, responsible for society-saving moonshot ideas. By the Makani team's estimate, there is enough wind energy circulating out there to power the entire world 100 times over. At present, they say, we are only capturing about 4 percent of that potential energy.[92] Researchers on the team therefore envisioned technology able to be positioned anywhere and deployed quickly and easily, even in areas normally considered off-limits for traditional wind installations. The team launched Makani Technologies under their parent company Alphabet to develop and test the potential for airborne wind energy technology.

What Makani's techs eventually created was essentially an aerodynamic wing system outfitted with rotors to fly in loops like kites do. Their carbon-fiber, lightweight kite is tethered to an energy storage system placed on the ground as the device soars at higher altitudes—including, as techs envisioned, areas where ground installations would be too cumbersome to build, and in disaster areas. Once airborne, the kite was guided by computer navigation using GPS tech and flight sensors. Every time the kite completed a loop, rotors on the wing would spin in the wind—powering the tech's on-board generators and producing electricity that was then dispatched along a tether to the ground.

Makani took advantage not only of the increased airflow created by higher altitudes, but the added force created by moving within wind gusts. The airflow experienced by a moving apparatus is apparently many times faster than for stationary objects. The wind speed differential reportedly allowed Makani to produce up to 600 kilowatts of electricity, which the team calculated could power 300 homes.

Sadly, in February 2020 Alphabet would, for reasons unknown, pull the plug on supporting Makani. Alphabet had been sinking billions into developing moonshot bets and the calls for cutting losses were becoming more numerous. It is altogether possible that while the technology was feasible, other renewables had simply become too competitive. "Despite strong technical progress," company CEO Fort Felker stated, "the road to commercialization is longer and riskier than hoped." So while cost metrics may have played the strongest role, a variety of growing calls for specific levels of marketability may have surpassed all others.

Netherlands-based Ampyx Power also uses a tethered drone energy system. But unlike Makani, the Ampyx team counts on wind to tug on the tether and activate a ground-based generator in the process. Another company, Germany-based KiteSwarms, uses a swarm of these same kinds of drones to generate amplified wind energy.

Curiously, researchers at the University of Manchester took the concept of wind-swept tethers generating electricity to test out renewable energy flags, which would normally just be waving in the wind. The team found that with the right upgrades, traditional flags can also harvest wind energy.[93]

In terms of other higher-altitude wind techs, reaching higher elevation courtesy of helium takes some of the energy expenditure out of staying airborne. And then, just as wave energy tech works by virtue of a river's force pushing water against turbine blades, a tethered helium-powered wind tech apparatus can essentially hover and collect energy.

Somerville-based Altaeros had, at one point, commercialized cylindrical blimps operating at roughly 2,000 feet above the ground with turbine tech housed in the center. These days, the company is more about renewably powered, high-altitude internet connectivity "towers" for urban and disaster-recovering communities—which is still a tech worth mentioning in the future of sustainable living.

Now, wind speeds are obviously very contextual. But even calculating regional averages in wind speeds on and offshore, low altitude and high altitude—offshore plus high altitude tends to produce the most wind power. German wind energy company SkySail estimates that with on-shore wind power at roughly 700 W/m^2, being offshore over the North Sea adds about 300 additional W/m^2. Likewise, being onshore with high altitude can provide roughly 1,500 W/m^2 while offshore at high altitude raises to roughly 1,800 W/m^2.[94] The company therefore uses easily deployable, offshore, tethered parachute-style tech to harness the energy of high altitude marine wind power. As the tethered kite swirls and loops, it eventually gains altitude. As it rises, energy gained will retract the unit to start over while excess energy is fed into the generator. Ultimately, the company boasts savings of up to 90 percent by using less material to harness faster wind speeds for more energy. SkySail is now joined in their energy explorations by similarly named AWE ventures TwingTec, EnerKite, Kitemill, and Windlift.

Wind energy is continuing to grow. According to the International Renewable Energy Agency (IRENA), wind energy is poised to account for more than a third of the entire world's energy needs at roughly 35 percent.[95] And yet, sadly, while misinformation campaigns exist against all forms of renewable energy—wind energy seems to have sustained the most manufactured rage of all. As a result, powerful and well-funded groups have launched initiatives to discredit wind in most of its forms, often by speculating about the health effects correlated to living near a wind turbine.

It's not like scientists aren't taking health concerns seriously: researchers have independently investigated any potential health impacts of living by a turbine for issues like nausea, sleep loss, and headaches. While researchers did find a few cases, they reportedly found no statistically significant correlation between health problems and living nearby to wind energy equipment.[96]

As more of the public learns about how wind power operates, the less likely they may be to fall for lies about our televisions going out when the wind stops blowing or about cancer being caused by turbine hum. In America, for example, large networks of grid operators routinely monitor power sources to augment falling sources with strong ones, typically with fail-safe options in place regardless. Their power to intervene in wind shortages will increase as interconnected data networks exchange incredible amounts of weather information in real time, empowering wind tech to develop further.

Incredibly, one study published in *Nature Climate Change* predicted that global wind speeds will actually pick up over the next several years until at least 2024 by an estimated 37 percent—making wind energy more profitable. A phenomenon called "global terrestrial stilling" has altered wind flow rates on average around the world, with wind speeds in 2017 providing approximately 17 percent more energy for turbines than in

2010.[97] Scientists expect the trend to continue for about ten more years, though emerging research into localized wind speeds will help determine that extent (and help determine new placements for turbines).

In the future, grid power allocation will be aided by advanced AI systems capable of detecting, via remote sensor, even slight changes in wind weather patterns far away from spinning turbines. Sensors tracking everything from cloud formations blocking sunlight to the impending wind speeds approaching turbines can all contribute toward stabilizing grid function. And, of course, to reduce the arguments often leveled against wind energy tech.

Wireless charging pad technology for electric vehicles. See page 28.
CREDIT: WITRICITY

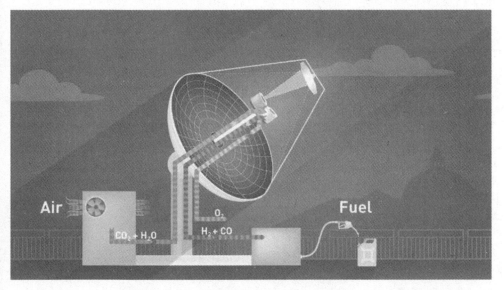

A "solar fuel" reactor designed to produce carbon neutral fuel from sunlight and air. See page 82. CREDIT: ETH ZURICH

Heliogen's concentrated solar power facility in Lancaster, California. See page 86. CREDIT: HELIOGEN

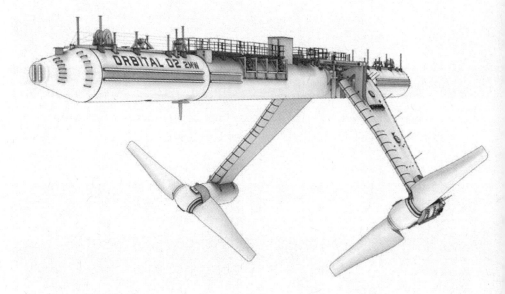

Equipment designed to harvest tidal energy. The main portion floats on the water's surface while the "arms" extend below. See page 93.
CREDIT: ORBITAL MARINE POWER

Nuclear fusion prototype. See page 114. CREDIT: GENERAL FUSION

CHAPTER 4

BROADENING OUR ENERGY PORTFOLIO

■

At 2°C, the world will start to look very different, which
is why politicians revised the warming cap down to 1.5°C
to make it a harder limit. But drawing artificial lines in
the sand is not really how the world works. It's not like
everything is ok at 1.99°C and at 2°C we're all going to die.
The answer is we need to stop doing what we've been
doing as quickly as we can. We've already crossed the line.

—JONATHAN FOLEY, PROJECT DRAWDOWN (2019)[1]

On September 12, 2019, scientists working at Stanford and the University of California, Los Angeles, announced that they had recently developed a new rooftop panel energy technology—for generating power from . . .

The night sky.[2]

The team began by envisioning a solar-panel-esque technology that would generate power on rooftops overnight, after the sun goes down. They had found that because our planet releases absorbed heat from the daytime back out into the cold of space, thermal heat energy would essentially be traveling upwards, in the opposite direction it normally does. So they flipped the script on thermal energy. The team placed a small thermoelectric generator between strips of insulation angled to gather heat passing up toward the sky, harnessing small amounts on the way. Their design, almost something of a solar panel in reverse, was taking advantage of an effect physicists call *radiative cooling*.

Radiative cooling is a process in which a sky-facing surface will pass off a portion of its heat to the atmosphere more readily as thermal radiation. As a result, surfaces will actually be slightly colder than the surrounding air. Which means that even if our ambient air outside is just above freezing, a surface may dip slightly below that level. We typically see the radiative cooling effect as frost present on the grass outside even after an above-freezing night.

Stanford and UCLA researchers stopped short of sensationalizing their new device at the time, as it was only able to generate enough electricity to power one LED light. However, the team also predicted that with improved engineering and R&D efforts, the device could be scaled up to about to twenty times the power—which still isn't necessarily a game-changer. But if we're picturing an inexpensive, easy add-on for your home, potentially powering up to twenty LED lights per night and eliminating that cost from your monthly bill forever—particularly if you're not even powering lights and are simply banking that excess energy via the in-home energy storage units we'll discuss later—it suddenly sounds like a technology worth discussing. Further, the team's device can even be built with easy-to-procure, off-the-shelf parts (see the citations for a link to the work in case you'd like to build one for yourself).

A scientist at the University of California, Davis, more recently developed what has been dubbed an "anti-solar cell" in offering precisely the same application—a rooftop panel able to generate power overnight. The device operates on a similar thermoelectric principle, but unlike Stanford's device can offer roughly 25 percent of the energy harnessed by a typical solar cell.[3] That's a solid improvement over the zero percent we're currently harnessing overnight with conventional panels.

I'm discussing a new nighttime energy application here not just because this is a book about next-gen renewable energy technology. I'm discussing it to show how human ingenuity never sleeps—and this matters more for a larger conversation about how we will realistically pull ourselves out of this mess we're in regarding our energy woes and our climate.

We now live on a planet where advanced research teams are working around the clock, all over the world. Somewhere out there—even now, as you read this sentence—research teams with incredible skills and knowledge about advanced sciences are working on developing new applications for renewable energy. Students are graduating from universities and technical colleges all over the world, applying their new skills and knowledge atop the knowledge base of what came before them. Radiative cooling energy tech is just one new application allowing us to harness yet more renewable energy. There are many others. And all of them will help us transition by baby steps or leaps toward the level of sustainability that we need to reach.

Some emerging developments have the power to not only revolutionize how we keep the lights on and heat our homes, but extend far beyond energy into many other aspects of our lives. A renewable energy source like sustainable biofuels, for example, is not new—but has enjoyed so many game-changing developments lately that biotechnology is now becoming one of the largest industries on the planet. Indeed, the biotech that scientists now use to obtain fuel is also allowing them to develop plastics, electronics, building materials, clothing, food, pharmaceuticals, and more. Much of it is carbon neutral and some of it is even carbon *negative*.

The speed of progress specifically in biotechnology has been astounding. Scientists are developing fuel from cultured algae that we can use to drive gas-powered vehicles. They're developing carbon-neutral jet fuel from landfill waste. They're engineering organisms to clean up our wastewaters while producing energy for us in the process. These milestones are now occurring so quickly that they often appear as brief headlines we may catch—as we get on with our lives outside of social media—before those headlines are replaced with newer news.

So let's get up to speed on where scientists are in terms of powering our world with biofuels. Let's catch up on how scientists are literally producing drop-in ready fuel from only sunlight, carbon dioxide, and other molecules in ambient air. Let's discuss the many incredible new energy developments being made around the world, all the new techs big and small, cumulatively bringing us closer to a sustainable world.

THE RISE AND FALL AND SUBSEQUENT
RISE OF BIOFUEL

By the time this book heads to press, you'll probably have seen those television ads being aired by fossil fuel energy giants featuring the incredible energy potential of algae. Yes, *algae*: that clingy seaweed you often hate to find creeping too close to the shoreline. Oil giants ExxonMobil and BP were among the first companies to begin running these ads in primetime, touting the potential for algae-based fuels to power not only our cars and trucks—but even our boats and planes. And for once, Big Energy is apparently not lying to us. It very much appears that algae and similar microorganisms will eventually accomplish these feats. It hasn't been cheap for Big Energy—and love it or hate it that previous planet-ruiners are now developing successful clean energy—but advanced new biofuels stand a solid chance of saving us from ourselves.

So what are "biofuels" and how might they revolutionize the transport and energy sectors?

In short, biofuels are basically any fuels we can derive from organic matter. The so-called "biomass" produced within organic material like trees and plants is often energy-rich—energy we can then burn directly or extract and refine to produce different kinds of fuel.

Traditional biofuel producers will grow large fields of fuel crops like wheat, corn, and sugarcane. Those crops are then harvested, and their sugars are typically used in a fermentation process involving microbes like *Escherichia coli* (E. coli) or *Saccharomyces cerevisiae* (brewer's yeast) to produce ethanol. The process is similar, in ways, to how store-bought alcohol is produced. Your favorite liquor comes from a particular type of fruit or crop, which is then fermented and distilled. Ethanol is combustible, of course, which is why nobody recommends trying to distill your own homemade liquor in the garage. But clever biofuel producers basically take this process and then refine ethanol into a blendable mix of combustible material that can be used as car, boat, and even jet fuel.

The primary benefit for biofuel producers is in crops' relative greater sustainability than finite fossil fuels. Crops can grow year after year, largely by themselves. With biofuel, there's no offshore oil rigs drilling deep into underwater crevices to obtain limited amounts of fuel. It's all just out in a field somewhere, growing on your plantation. Better yet, though: because crops are building biomass through the process of photosynthesis, they are also absorbing carbon dioxide for their growth. The CO_2 that crops absorb stays with them throughout their lifecycle leading up to harvest. And while that carbon is eventually returned to the atmosphere when the fuel is burned, the fuel is said to be "carbon neutral" because only the amount absorbed during growth is later released.

The secondary draw is that biofuels can be comparatively cheap. Running that offshore oil rig and transporting fossil fuel is, for a variety of reasons, more expensive and troublesome than simply harvesting fields of select crops—to say nothing of the impending legal fees you may be facing if you happen to run a fossil fuel company (or the tax fees you will inevitably be paying for your company's carbon emissions in years to come). What you therefore get with biofuels is a comparatively cheaper, more sustainable fuel that's also lighter on the carbon emissions.

More importantly yet, biofuel producers may one day be able to make much of the process carbon negative. Which is to say, a process absorbing more CO_2 than the amount it emits. Immensely important for remedying climate change and repairing our planet, to say the least.

If we take biofuel crops that have absorbed carbon dioxide to grow and then capture the carbon released when they are burned for fuel, the net amount of atmospheric CO_2 will be less than before. Whether we burn crops directly for bioenergy or refine their sugars for fermentation and process the material into ethanol, capturing the carbon thereafter will result in a net loss of CO_2. The process of combining bioenergy with capture and sequestration (BECCS) is a concept we will explore in chapter nine, which covers the many strategies we have for actually addressing climate change directly. But for now, that's just one more benefit for potential biofuel producers to pursue.

Cultivating a vast plantation of biofuel crops can come with its own difficulties, however. A lot of water, agricultural upkeep, and attention must be paid in refining these materials. And not all crops offer the same payoff in terms of energy obtained. So prospective biofuel producers have a vested interest in finding the most superior fuel-producing crops and reducing the waste before going all-in.

Capitalism being what it is, these concerns eventually led enterprising biofuel scientists to consider new alternatives. Enter algae and cyanobacteria—other photosynthetic organisms that, like fields of crops, can grow via sunlight, capture carbon, and ultimately produce fuels for us. Except they don't require huge fields, tons of wasted water, and pest-control. In fact, unlike crops, algae and cyanobacteria can thrive just sitting in big vats of water out in the sun regardless of soil or climate conditions.

Scientists working for Big Energy have actually been experimenting with algae for decades. Essentially every time the price of oil skyrockets due to shortages or warfare, scientists get a flush of funding to make this product possible. Or, of course, during times when public opinion becomes wholly opposed to their industry's business model—which is where we find ourselves now.

The problem in developing algal biofuels has consistently been that it is never as powerful as it needs to be. As a result, the PR pumped into algae back when it was in its earliest (unsuccessful) stages resulted in bursting bubbles and loss of funding. But the situation has slowly been changing. New research and new technology have increasingly led energy scientists toward commercially viable algae-based biofuel. They have spent decades studying which algal characteristics produce superior fuels—scouring the world for successful strains, seeking out greater biomass, identifying superior "lipid reserves" in some strains over others. Faster growth rates. Easier extraction rates. Cheaper drying methods. Faster fuel conversion rates. Cheaper, more powerful biofuel as a result.

Biofuel scientists have laboriously studied the genetic components of traits in algae strains beneficial for harvesting biofuels. They have "bioprospected" algae strains from around the world in hopes of cross-breeding these traits into hybridized super-populations. They have looked for successful mutations driving these populations forward. But attempting to create superior groups of algae and cyanobacteria through natural means has not always yielded very compelling results—which is to say that traditional Darwinian evolution has been slow and cumbersome for producing super-algae. So scientists have now turned to new techniques in genetic research.

Incredible new advancements in genome sequencing, bioinformatics, and precise gene editing technologies have opened up vast new possibilities for biofuel scientists. Newly developed genetics techniques like RNA interference (RNAi), CRISPR-Cas9, and the even newer "prime editing" technique have provided researchers with amazing new opportunities to achieve desirable genetic outcomes—super-algae for fuels, along with any other organism they might be studying.[4]

At its most basic, CRISPR is a new technology offering us an unparalleled ability to edit genetic material on a rapid scale. The acronym itself stands for "clustered regularly interspaced short palindromic repeats"—which is a reference to the repeating sequences of DNA often known for their use by the immune system.

Consider the defense system available to bacteria. When a bacterium uses enzymes to destroy an invading virus, some of the virus's genetic code is snipped and stored for reference in CRISPR spaces. Should a bacterium encounter that same invader again, it can use the stored genetic information essentially like a database of viral fingerprints to identify the problem sooner.

In a scientific breakthrough we'll read about in our history books for the rest of time, researchers recognized that an enzyme called Cas9 could be fed artificial RNA (rather like providing it a fake fingerprint to assess). Scientists found that doing so would be a diligent internal search for that genetic code. What happens next has been the subject of technological progress over time. But originally, CRISPR tech would essentially find pieces of a target genetic code and either alter or shut them down. At a more detached level, the end result was a newfound ability to tailor enzymes that seek out and alter parts of an organism's genetic code. Even newer innovations on this process improve it even further, offering bioengineers what is basically the search-and-replace function you use in word processing software like Microsoft Word. Except, searching for and replacing genetic sequences.

This might sound frightening, depending on how responsible you believe we can be with the technology. CRISPR and prime editing now offer us the power to direct evolution ourselves. We can empower advantageous traits or reduce negative ones. We even have the power to edit the so-called "germline" of organisms—cells that pass on traits during reproduction, creating a heritable condition. Doing so will create heritable conditions that are tough to undo, so we realistically stand at a precipice of power—wondering whether we ought to be using it on ourselves. These discoveries have naturally led to discussions of concepts like "designer babies" and engineering positive traits into newborns who will then pass on those traits to future generations. But for better or worse, we are not quite that sophisticated yet. So there's still time to have these important conversations about how we might use our newfound abilities.

Using CRISPR, researchers have been able to identify the transcriptional regulators (read: genetic codes) of traits within algae groups advantageous for biofuel production.[5] For any genetics buffs reading this, new strategies for altering algae or cyanobacteria in advantageous ways have included advancements in ribosome binding sites, genetic promoters, riboswitches, modular vector systems, markerless selection systems, and more. That is all beyond the scope of this book—but the bottom line for now is that we are using advanced genetics editing tools to snip and splice the most advantageous biological traits we can find together into algal colonies and cyanobacteria while simultaneously diminishing the negative. And we're quite likely to continue doing this well into the future.

More incredibly yet, scientists are now exploring methods of artificially increasing beneficial traits in various sources of biofuel through the emerging field of *synthetic biology*. Synthetic biology is an umbrella term for a multidisciplinary new field of science geared toward the rational creation of new biological parts—or the redesigning of pre-existing evolutionary blueprints. "Synbio" for short, the expanding new knowledge-base of synthetic biology is increasingly allowing bioengineers to tailor-make particular genetic components. As a result, several microorganisms are now essentially becoming tiny cellular factories we've built—crafted with the necessary internal components needed to scaffold certain sets of chemicals into others.

The game-changing development in energy is that corporate giants have now partnered with research teams capable of pursuing these ends. Companies like BP and Exxon have recently been able to artificially increase the capabilities of algal cultures even beyond bioprospecting. They have raised energy-dense lipid production, carbohydrate production, and even have newer harvest-free capabilities they may be able to bring to scale. Given that energy companies are now broadcasting their power to do so on primetime television, I suspect that they're waiting on a massive scale-up before roll-out a few years to a decade down the line.

How does this look in real life? Well, rather than massive oil refineries, energy producers will have tubes of water teeming with tiny green life or batch ponds in a facility. The algae or cyanobacteria will lap up sunlight and CO_2 to grow. Engineers will periodically drain and filter these tubes and ponds to obtain fully developed biomass. These components are then harvested from the mix in a process of cell harvesting, drying, and lipid extraction to refine into the carbon-neutral biofuels we need.

It's important that we keep our expectations in check, of course. Biofuels have been criticized for their supposed inability to provide enough fuel for the energy spent in processing. Critics often run cost metric analyses through assessments of currently existing technology, however, without much consideration for how new developments will shave costs from various parts of the process. And reasonably so, given that predicting the future is clearly very difficult. But it is therefore equally important that we appreciate when new innovations have pushed the needle more toward commercial viability.

Throughout 2019 alone, scientists reached several remarkable milestones for biofuel refinement. One new technique was developed for separating lipids in one step (and in seconds, no less) using a specialized jet mixer apparatus—and another new technique utilized reusable nanoporous filtering materials for increased drying speeds at a vastly reduced cost. Still another promising new development came from scientists working in synthetic biology, literally designing a more efficient metabolic pathway for photosynthetic organisms harvesting solar energy. This latter development was particularly promising, given that critics sometimes cite the "natural ceiling" of biomass production in algae and plantation crops as a crucial limitation.

Newer innovations on using sunlight to power up organisms for fuel will be something we return to later in this chapter (and that last part about designing new metabolic pathways will be expanded upon further in the book). For now, even before algal biofuels have been widely utilized in the real world, another approach called sun-to-fuel has also been developed with similar equipment in mind. Incredibly, because the processes of algal cell harvesting, drying, and lipid extraction can still be expensive and lengthy—intrepid new scientists have developed genetically engineered strains of cyanobacteria able to produce fuel *directly*. They still grow with only sunlight and CO_2 in water, but they now can simply absorb solar energy and produce fuel in the process, bypassing many other steps.

When you think about microorganisms as tiny cellular factories, fuel is just one of the many things they could potentially make. And as it turns out, engineered algae and cyanobacteria are now capable of metabolizing sunlight, CO_2, and water into all kinds of things. Another upcoming chapter will discuss all the cool things modified algae and cyanobacteria are now making for us.

In the case of fuel, cyanobacteria lap up sunlight and excrete metabolized fuel components directly into the bioreactor in which they live. Scientists then separate out those fuel components from the vat in a continuous filtering process. The EU's "Photofuel" project is among the growing number of international research groups spearheading this fascinating new work.[6] The team has handy web-graphics available online to help understand how the process works, if you're interested in a more visual assessment. The graphics feature a batch of engineered cyanobacteria in a vat with an attached gas pump nozzle adjacent to filter out the fuel produced, which is butanol (even if a real life version won't be quite so simple). Butanol is a relatively green, fourth-generation biofuel that can be used in the automotive industry as a drop-in fuel substitute for gasoline. But for the moment, sun-to-fuel tech is still further off than the now "traditional" means of harvesting algal biofuel from biomass.

Ultimately, then, research into utilizing photosynthetic life to produce biofuel has progressed along a remarkable trajectory. First, attempting to prospect for superior strains showcasing superior advantages in different aspects of production. Then, either cross-breeding or bioengineering desirable traits that exist in different species into a hybridized species capable of doing an all-in-one process. Then, further to that: working to develop novel genetic traits that would potentially never have appeared throughout the normal course of evolution, which is not nearly as scary as it might sound.

More incredibly yet, some research teams still see much more opportunity in algal biofuels. Because they require fewer resources and capture carbon for us, it is largely just the fuel we want from algal colonies. But if strains can perform multiple other feats while doing so, that will just be a shot in the arm for a growing industry of cleaner fuel. So any time researchers can produce strains capable of these feats, it's something to celebrate for a more sustainable world.

For example, the process discussed so far has involved either batch ponds or tubes filled with microscopic algae in a refinement facility somewhere. What if those algae were instead living in a batch of wastewater produced from industrial manufacturing? Could they still live and produce fuel, to save ourselves from using mass amounts of new water? Or, in the case of algae strains engineered to *love* wastewater—could they not only thrive, but break down the harmful chemicals in wastewater while simultaneously producing fuel? As it turns out, they can.

Consider the work of a research team in India prospecting a strain of micro-algae capable of both producing biodiesel and remediating wastewater.[7] The alga involved was engineered to target the heavy metals in wastewater, ultimately removing roughly 87 percent of a sample's arsenic content and about 95 percent of its cadmium—as well as mitigating high percentages of chlorine, fluorine, and bromide, all while still accumulating the necessary lipids and carbohydrates that scientists can later convert into bioethanol and biodiesel.

Additional prospects for biofuel-producing microorganisms will increase as scientists highlight the genetic components they need and master the introduction of others. Some of these organisms may potentially (or perhaps inevitably) seem rather unnatural in light of how traditional algae looks and functions.. And there's certainly a larger debate to be had about the extent to which humanity should go meddling in the genetic material of other species. Though on the bioengineering side, we've already been dictating the evolution of many other species up till now regardless. Some of our favorite pets, of course, come from a long line of directed evolution. Some of our favorite fruits have been carefully directed over generations to favor some genes and traits over others. Technology like CRISPR can simply speed up this process.

Now, as I noted earlier, some scientists are also working on developing fuels using sunlight and CO_2 entirely without photosynthetic life. This is because we have progressively developed a technique called "artificial photosynthesis" to harness plants' natural power for ourselves. And why not? Photosynthetic organisms do not hold a monopoly on this chemical transaction. We, too, can harness the power of sunlight to spur similar reactions using only technology. Before we get there, however, it will be crucial to understand another one of our new potential energy outlets: hydrogen fuel. Let's discuss hydrogen briefly before jumping into sun-to-liquid fuels.

THE HYDROGEN ECONOMY: H_2 FUEL

Consider that hydrogen is the most abundant element in the universe. Next, consider that hydrogen can actually be used as a zero-emissions fuel and that the only by-product left over from using that fuel is pure, clean, drinkable water. Imagine a plentiful resource able to provide sustainable energy to power vehicles or even whole buildings with little to no threat for the environment during use. That is the promise of hydrogen fuel. And while the hydrogen option may be a bit more complicated when we consider everything that goes into producing H_2 as usable fuel, we do have plenty of compelling evidence indicating that it will play a pivotal role in our transition away from fossil fuels.

Many of us know about hydrogen fuel thanks to so-called "fuel cell" cars available on the market already. Vehicles like Toyota's Mirai model can run from an emissions-free hydrogen fuel cell, producing water from its exhaust that you can even drink after. The car can fill up at designated gas stations with a hydrogen fuel pump, complete with your typical nozzle and pay-at-the-pump convenience.

So why haven't we expanded with this technology?

In reality, using hydrogen for energy has been discussed for almost two centuries. Hydrogen fuel cells were first invented about 180 years ago by William Grove, an amateur Welsh chemist studying gaseous voltaics in the nineteenth century. The basic structure of his early prototypes—combining hydrogen and oxygen with electrodes and electrolytes (like a battery) to produce electricity and heat—is essentially what we do today, albeit with a few variations in place of the liquid electrolytes.

The primary catch in using H_2 for fuel is that while hydrogen is abundant, it does not often appear in isolation. Hydrogen atoms are typically attached to others in various molecules, requiring a costly process of separating the hydrogen from whatever it's bonded to before we can potentially use it for fuel. We can take a molecule like water's H_2O, for example, and separate it into H_2 and O, leaving us with hydrogen and oxygen thereafter. We normally use an energy-intensive, carbon-heavy process of steam reformation to accomplish this—or the costly process of electrolysis in splitting water with electricity. Which means that the cost for completing these operations has always been debilitatingly high. To say nothing of the subsequent storage and transport costs in either housing the newly compressed hydrogen gas or freezing it into a cryogenic liquid for tankers to carry to fuel stations.

The off-set in energy spent for its return is currently not in hydrogen's favor for vehicles. Several auto-makers have therefore moved toward battery-powered electric vehicles (BEVs) rather than hydrogen EVs. But the promise of running whole buildings and essentially anything else we desire on limitless, clean energy is still a very alluring concept.

Most progress in obtaining usable hydrogen has come from improving electrolysis—in terms of superior catalysts to drive the process forward and in reducing the amount of energy involved. So-called "green hydrogen" is being produced by *renewably powered electrolysis*. This process is variously referred to as artificial photosynthesis and is potentially hydrogen's best shot at being a civilizational staple.

Artificial photosynthesis is functionally similar to general photosynthesis. Normally, plants will take the energy of sunlight and split water it has absorbed into hydrogen and oxygen. The oxygen is later released from the plant, providing us with breathable air, while the hydrogen is further utilized throughout the later parts of photosynthesis. Incredibly, just like any good factory will separate dangerous materials in different areas, plants had long ago adapted cellular mechanisms to separate the hydrogen and oxygen during extraction—to spare themselves the spontaneous combustion of mixing hydrogen and oxygen in the presence of heat.

Artificial photosynthesis begins by running an electric current through water. This process, electrolysis, will separate water into its components that we then separate. And at this point, it's worth remembering that this is the way artificial photosynthesis *begins*. How it proceeds from there will be a point continually revisited throughout this book. Because while we can then shuttle away the hydrogen and isolate it for later use in fuel, enterprising scientists have also discovered other things we can do with spare H_2.

For now, in terms of harnessing hydrogen for use in fuel cells, the process involves separating hydrogen and then storing it. Harvard professor of energy science Daniel Nocera was not the first scientist to accomplish this feat, but was perhaps among the most famous for popularizing it. His "artificial leaf" technology in 2008 essentially used sunlight to power the electrolysis and provide hydrogen—for the purpose of storing and using it later. Nocera's design made headlines for using a PV panel to split the H_2O (while the sun was shining) and then reserving the hydrogen and oxygen for later combination to produce energy. The result? In theory, renewable energy powering your home even after the sun goes down.

The concept of powering up electrolysis devices with "excess" renewable energy during times of plenty and saving that power for later is something we will also return to again later on. Because at a larger level, we sometimes have way more renewable energy than we need at peak hours. Finding a productive use for that excess energy will go a long way toward making full-scale renewable energy workable during non-peak hours. And in terms of hydrogen specifically, researchers have already noted that in places like Germany and Texas, where far more wind power is produced than grids need at particular times, the excess power can be economically utilized to produce hydrogen.[8] That hydrogen can then recover the excess energy later on via hydrogen-powered fuel cells. But I'm getting ahead of myself: let's continue with how hydrogen energy factors into fuel.

WHAT TO DO WITH H_2?

Hydrogen fuel is not without its critics. For one, the energy efficiency is currently inferior to using a simple battery-powered EV for transport. Tesla's Elon Musk once described the prospect of hydrogen electric cars as, "mind-bogglingly stupid" and referred to fuel cells as "fool cells."[9] Of course, Musk has a significant economic stake in how sustainable vehicle research is carried out.

Toyota has only sold about 5,000 Mirai models in the US since releasing the car back in 2015 (at this time), while Honda has sold even fewer Clarity models. And in a state like California, always on the forefront of decarbonizing, there are only 39 public hydrogen fueling stations at this time of writing (though 200 more are reportedly in the works by 2025). Other states like New York, Massachusetts, New Jersey, and Rhode Island have hydrogen fueling station plans in various stages of development. Here in Canada, Toyota and Honda have partnered up with the provincial government of Quebec to enable more hydrogen infrastructure in Montreal. So the work is in progress.

The Japanese government has invested heavily in developing hydrogen infrastructure and processing—which figures, because Japan is the world's largest importer of liquefied natural gas (LNG) and also in the top four importers of coal and oil. They would like to change this status going forward. Toyota, Nissan, Honda, and nine other Japanese companies have all joined forces to create the Japan H_2 Mobility (JHyM) program and create 80 new hydrogen fueling stations by 2022. Japan can also boast the world's first hotel partially powered by hydrogen, in Tokyo.

A bonus of developing hydrogen fuel cell technology (unlike battery-powered electric vehicles) is that companies will be able to use all the pre-existing gas infrastructure. If hydrogen is turned into a compressed gas or liquid, existing pipelines can be used—existing fuel tanks, existing service station storage chambers, etcetera.

A company like Germany-based Hydrogenious Technologies specializes in producing Liquid Organic Hydrogen Carriers (LOHC)—for transporting and storing hydrogen turned from a gas into a liquid. Technology able to accomplish this feat can far more easily replace existing gasoline infrastructure not just in the future—but almost immediately. Hydrogenious has been among the most commercially successful in solving both the "evaporative" challenges of gaseous hydrogen and the flammability issues.

Whichever company commercializes hydrogen fuel successfully, H_2 is first procured through either steam reforming or from renewably-powered electrolysis as a gas. Engineers then perform a process called hydrogenation: the chemical bonding of hydrogen molecules to a liquid carrier via catalytic reaction. As a liquid, hydrogen fuel is then transported via traditional transport infrastructures. In its liquid form, hydrogen is not nearly as flammable, is non-explosive, non-toxic, can accommodate a wide range of handling temperatures, and is not currently classified as a hazardous good (bypassing at least a few regulatory hurdles). Then, once on site, basically the reverse can happen: dehydrogenation and the release of hydrogen molecules from the liquid carrier medium via a catalytic reaction. Hydrogen is then able to be used as a fuel in vehicle fuel cells, a chemical for industrial processes, or as an energy storage medium in re-electrification via fuel cell.

Hydrogen research is still ramping up around the world. In Germany, the world's first hydrogen-powered trains offer a zero-emissions alternative to traditional diesel-run cars. Off the coast of Scotland, the European Marine Energy Centre has launched a project on the island of Eday to use tidal power to produce hydrogen. The hydrogen power system being developed there will eventually power car and passenger ferries in the surrounding islands. In the UK, the $H_2$1 North of England project aims to decarbonize 3.7 million UK homes and businesses by converting the gas grid to hydrogen produced from gas with carbon capture.[10] In France, one start-up company recently launched the world's first series of hydrogen-powered EV bicycles.[11]

Or how about hydrogen-powered drones? South Korean company MetaVista recently broke the standing record of longest flight time for a fuel cell UAV in the air. With longer flight times, hydrogen-powered drones potentially offer greater possibilities for drone-driven delivery systems, scientific data-gathering programs, and military reconnaissance. MetaVista even claims that the effect of using liquid H_2 brings an increase of three times longer flight spans for hydrogen-powered drones.[12]

So there are clear signs that hydrogen can be renewably obtained, reliably used, and safely stored. The wild cards are that researchers are only now developing greener and cheaper methods of producing hydrogen in general. Biohydrogen and thermochemical water-splitting are two new state-of-the-art methods just over the horizon that show massive promise for developing clean fuel. It is worth taking a brief detour of both, then, to assess how hydrogen may impact our energy going forward.

Biohydrogen is simply any hydrogen gleaned from biological sources. As with the production of biofuels, developing organisms to "produce" hydrogen is yet another case of just building the right cellular factory. And once again, biologists hoping to solve complex problems first look to nature for solutions hoping to build on what they find.

As it turns out, there are plenty of microorganisms out there that can grow without oxygen. Resilient microbes will often use whatever they can find within an environment to survive—which sometimes means consuming hydrogen. Strains of E. coli will consume hydrogen in various circumstances, leading synthetic biologists to believe that this process can be reverse-engineered. But there are also other organisms out there, like algae, that can already produce hydrogen naturally with a hydrogenase enzyme in special circumstances. In fact, algal hydrogenases are among the most efficient H_2-generating biocatalysts out there.[13] As a result, there are biological means for mass-producing hydrogen in theory. Our research teams just have to work out the details.

Incredibly, one research team led by scientists at Arizona State University recently announced they had rewired the photosynthetic arrangement of a green algae strain to produce biohydrogen more directly. The team created a photosynthetic chimera strain able to divert a portion of the electrons normally reserved for water-splitting to create molecular hydrogen instead. Cells containing the rewired photosynthetic structure were actually able to generate hydrogen at high rates (dependent upon the presence of light, of course). Research here is still ongoing, but synthetic biology may eventually save the day for sustainable hydrogen production.

The prospect of having microbes or algae cultures producing biohydrogen is a tantalizing concept because it's cheap and can be done anywhere. Scientists are still working on finding hydrogenases and studying how best to obtain sufficient amounts of active hydrogenase enzyme. When that happens, though—because bacteria like E. coli are much easier to engineer than algae strains—this genetic capability may then be hybridized into engineered bacteria. What seems potentially plausible, then, is an engineered hybrid organism sharing genetic capabilities from both species. Should that happen, economics around fuel cells will have changed completely.

Perhaps even more compelling is the use of concentrated solar power (CSP) and thermochemical water-splitting. This is a concept we'll explore more at the end of this chapter, but it is essentially possible to heat water molecules up to a point at which they separate. We can potentially direct solar-powered heat in excess of water's splitting point, giving us another potential avenue for generating large amounts of hydrogen cleanly.

COMBINING SUNLIGHT, BIOFUEL, AND HYDROGEN: SUN-TO-LIQUID FUEL

This is where the biofuel and hydrogen stories really start to pick up.

I noted that scientists are utilizing artificial photosynthesis for the purpose of isolating hydrogen. But some teams are integrating a bionic component involving bacterial (but still not photosynthetic) life into their design. In these cases, solar-powered electrolysis begins as it normally would, splitting water into hydrogen and oxygen. But after the H_2 and O have been separated, the hydrogen is then shuttled to bacteria engineered to consume it. A genetically modified *Xanthobacter autotrophicus* bacteria will consume the hydrogen and inhale CO_2, producing simple fuels in the process. We can then refine these materials into other, more usable fuels (like diesel, gasoline, and jet fuel). Or, we can refine the product into additional chemicals like fertilizers and bioplastics. Furthermore, we can even drive this process with entirely renewable energy sources. That, in a nutshell, is the sun-to-liquid fuel promise of the future.

Because scientists need continual supplies of water to power artificial photosynthesis, teams from around the world have been experimenting with different designs with ready access to H_2O. Some teams have been working on splitting seawater, others have used wastewater, still others have simply been using the moisture in ambient air with a solar reactor. All of these methods are considerably greener than how we obtain hydrogen at present and, if used widely, could produce carbon neutral fuel at mass scales.

Research out of Norway and Switzerland published in PNAS, for example, has described how we can use renewable energy from solar PV tech and offshore wind turbines to drive the electrolysis of seawater. We can then take the hydrogen produced—along with the CO_2 in seawater—to make methanol, a fuel we can use in diesel engines or refine for traditional combustion engines.[14] Vast swathes of ocean off the coasts of South America, the Arabian Gulf, Southeast Asia, and North Australia are comparatively free from large waves and extreme weather. They therefore make prime candidates for such installations. Researchers estimated that just one half-square-mile floating solar farm of 70 panels could reportedly produce over 15,000 tons of methanol per year (roughly about the fuel of one Boeing 737 making 300 round-trip flights between New York City and Phoenix[15]).

What makes the development particularly interesting is that engineers may one day utilize this technology to separate hydrogen from seawater in order to isolate the O from H_2O—in the interest of producing breathable oxygen under water. Underwater tidal-based renewable energy sources, which we'll get to in the next chapter, could help fuel that process.

Closer to home, scientists at Stanford have developed a method of generating hydrogen fuel using solar power and the saltwater of San Francisco Bay.[16] Another research team at Princeton has focused on using wastewater from a local brewery. Their trick? Bacteria feeding on the organic matter in wastewater can generate the requisite electrical current driving part of the electrolysis.[17] The improvement helped Princeton's team run the process for days, which is currently something today's designs lack as of early 2020.

Alternatively, we could harvest the moisture in ambient air and split that water with captured CO_2. We now have the technology to create fuel directly from sunlight and air, usable by current transport vehicles. In what should have been a headline to break the internet, researchers from ETH Zurich (essentially the Swiss answer to MIT) reported in June 2019 that they were able to produce, and I quote: "carbon-neutral fuel made from sunlight and air" with miniature solar refinery systems.[18] In fact, the team not only demonstrated a capability of producing fuel from sunlight and air, they did so in real-world test conditions on the roof of their lab. Other units, they say, could be placed essentially anywhere. All while producing a fuel that emits the same level of CO_2 extracted to produce it.

Under Zurich's "sun-to-liquid" model, the water and CO_2 are ultimately extracted directly from ambient air via adsorption. The H_2O and CO_2

molecules are then split using solar energy in a similar manner, although in this case a solar reactor is used to generate higher temperatures. The product of mixing the subsequent hydrogen and carbon monoxide is syngas, which is then liquefied into hydrocarbon fuel like methanol or kerosene.

The research team at Zurich currently produces very little fuel directly: less than a decilitre per day. But they believe they can ramp it up to about 10 million litres of methanol per year by 2025. How all this might work is a tandem model like that used by Swiss companies Climeworks and Synhelion. Climeworks and Synhelion essentially compartmentalize this process: Climeworks captures the CO_2 and Synhelion aims to produce economically viable CO_2-neutral drop-in fuels, entirely compatible with the current global fuel infrastructure. Which, in terms of a difficult-to-decarbonize industry like aviation, will be game-changing. Zurich's researchers estimated that one solar refinery plant roughly the size of a square kilometer could produce about 20,000 litres of more sustainable cleaner kerosene per day.[19] Or, if we're dreaming bigger (and I'd like to think we are): a plant roughly a third the size of California's Mojave Desert could reportedly cover the fuel requirements of the entire aviation industry.[20]

Of course, Synhelion operates independently as well—as a "solar fuels" production company, the spin-off from ETH Zurich also utilizes a concentrated solar plant with high-temperature thermal heat-absorbing materials to produce drop-in capable fuels more efficiently than going through biofuels *or* a power to liquid model. Which, again, will continually cause us to re-think just how feasible dropping our transport emissions will be. Refer to page 61 in the photo section preceding this chapter for an image of how Synhelion's solar fuel design works.

Ultimately, there could be a passionate back-and-forth between EV fans and those who believe biofuels can save conventional cars. Because if a biofuel is sustainably harvested and burns carbon neutral, then there's nothing inherently wrong with using a combustion engine. It's always just been the nature of the fuel we use inside of them. So those who argue that EVs and their associated infrastructure are not necessary will have a few points to make. Then again, those who argue that vast biofuel plantations propping up a global transport sector may not be wise—given the insane amounts of land and water required—will also have a few points.

Estimates analyzing up-and-downstream greenhouse gas profile assessments have revealed that biofuel options can offer us between 40 and a stunning 96 percent greenhouse gas reductions over traditional vehicles.[21] Moreover, biofuels can benefit further from even newer advancements courtesy of a technology developed at the US Department of Energy's Oak Ridge National Laboratory and scaled up by Irvine-based Vertimass LLC. The team's Consolidated Dehydration and Oligomerization (CADO) technique offers a one-step (versus three) efficiency upgrade for turning ethanol into hydrocarbon blend stocks able to complement aviation, shipping, and other heavy-duty vehicle fuels. The upgrade has allowed bio-based ethanol to achieve price parity with conventional fuel while reportedly offering all the sustainability benefits. We should all be prepared, then, for new developments like this to change the gas vs. EV debate we're having—and that perhaps a good blend of both will suit our purposes in different circumstances.

THE MANY FACES OF ENERGY: INNOVATORS WILL INNOVATE

Most of the renewable and clean energy sources we've discussed are beginning to converge in mutually beneficial ways. Solar PV tech helps to spur artificial photosynthesis. Hydrogen fuel tech helps to drive sun-to-liquid tech. Biofuel tech helps both. But one of the most surprising aspects of researching renewable energy for this book has been the ways scientists are harnessing all new *sources* of energy. Like radiative cooling, for example. And as it turns out, scientists are beginning to capitalize on many other surprising sources of renewable energy.

For example, consider the discussion about energy harnessed from heat back in the section about solar thermal panel technology. That thermal energy is provided to us from the sun. But consider that we could just as easily use thermal energy from anywhere with the same technology—and that we waste an exorbitant amount of heat from all of our electronics. The result? A new renewable energy source capitalizing on *waste heat*.

It is estimated that as much as two-thirds of energy consumed in the US every year is simply wasted as heat.[22] The sources are essentially innumerable: exhaust from every vehicle on the road, your phone, your laptop or PC, the TV, and all the refrigerators out there running nonstop. Given what we now know about the potential of harnessing heat for thermal energy—just imagine if we could capture more of that heat to generate additional energy. Not surprisingly, new developments are in the works in labs around the world to do precisely this.

One team at the University of Hong Kong recently developed a new Direct Thermal Charging Cell (DTCC) technology, capable of efficiently converting lower levels of heat to electricity. The team specifically targeted "low grade heat" in the 80 to 150°C (176 to 302°F) range normally just given away to ambient air following industrial processes.[23] The device will reportedly be tested in HVAC (heating, ventilation, and air conditioning) systems to recycle low-grade heat into electricity for channeling into a central device. Or, the team believes their small, bendable, low-cost, stackable design (measuring only about an inch and a half squared) could be integrated into windowpanes for powering household devices. The equipment could also power portable devices wherever enough heat is present. Crucially, the team believes their tech could also help to power wearable electronic devices (or personal medical equipment) with the help of nothing more than body heat.

Powering wearable technology is a concept actively being explored in labs around the world. I briefly mentioned "smart clothes" (or "green wearables," as some prefer) in the introduction for this book, and the case seems to fit my point here—we nearly always need to have clothes on in public, so let's make them *do something* for us. That we consistently move about in physical space creates the potential for harnessing kinetic energy and sunlight any time we are outside. Yarn-like zinc ion threading woven into fabrics destined to be shirts has been demonstrated to generate electrical power whenever bent, stretched, or exposed to sunlight. A material like perovskite, which is flexible and "printable" like ink, is already one potential candidate for lining future smart clothes, provided we can protect the material from the air around us.

Scientists have also demonstrated successful trials with tiny, nano-sized generators we might fit into future smart clothes. Scientists at Rice University, for example, found that they could harness the so-called "triboelectric" effect when materials gather a charge through contact—potentially every time a heel strikes ground or when arms swing during walking. With flip flops for a test device, the team generated about 0.22 millijoules of energy during a 1-kilometer walk. Certainly not a lot of energy, but enough to power wearable health sensors.[24]

Smart clothing sensors could realistically help monitor pulse, breathing rate, blood pressure, joint movements, and more. Engineers from North

Carolina State University have crafted a thin, flexible device to monitor health sensors, albeit powered by body heat rather than triboelectric charges.[25] Scientists believe self-sustaining sensors like these could also detect the earlier stages of some health conditions, including dehydration or sores forming on immobilized patients.

Or consider the concept of concentrated solar power (CSP) mentioned earlier. The concept of CSP plants involves taking hundreds or even thousands of mirrors, placing them in a field and reflecting the sun's rays towards a single point. Motorized "heliostats" automatically rotate to track the sun as it moves across the sky to maintain the mirrors' angle of redirection. A tower is positioned in the middle of these thousands of mirrors to create a reference point and house a receiver at the top for collecting all the sunlight. As a result, CSP plants have typically reached temperatures of up to about 565°C. CSP plants efficiently reach this temperature easily enough that it is sufficient to store thermal energy and relay clean electricity after the sun goes down, which is something cities around the world use already.

The Noor-Ouarzazate concentrated solar power complex in Morocco, for example, produces enough electricity to power a city the size of Prague. The array takes up approximately 3,500 ft. in square footage, with a central cylinder full of molten salt to house thermal energy for up to three hours of power after sundown.[26] The TuNur project planned for Southwest Tunisia will likewise utilize CSP technology with a molten salt storage component spread out over 5,000 hectares. China has used six large-scale CSP installations leveraging the off-set of hot molten salt to continue providing power long after the sun goes down. We'll discuss this design in more detail in the next chapter, which deals with next-gen renewable energy storage for future grids to compete with fossil fuel reliability.

Aside from turning sunlight into electricity, concentrated solar power (CSP) plants can also utilize channeled heat for doing other useful work (freeing up the need for what may have been large amounts of energy to do that same work). New advancements in efficiently channeling sunlight are providing us more opportunities to augment industrial processes with cleaner CSP alternatives. Moreover, new materials capable of absorbing larger amounts of heat will also allow CSP plants to perform industrial feats that plants of the past could not.

Consider, for example, the concept of coupling new thermal heat-absorbing materials in CSP plants with the kinds of advancements in AI and machine learning we have discussed so far. What you end up with is a newfound ability for concentrated solar plants to track the sun's motion, channel more of its energy and ultimately reach higher temperatures than ever before in commercial settings. One notable new company pioneering precisely this new combination—and backed by investment heavyweights like Bill Gates and Patrick Soon-Shiong (owner of the *Los Angeles Times*)—is Pasadena, California-based Heliogen.

Heliogen's contributions to CSP design are noteworthy for several reasons. For one, the outfit is comprised of trail-blazing scientists largely from American institutional powerhouses like Caltech and MIT. But more to the point, Heliogen combines artificial intelligence and advanced vision software to hyper-accurately align every mirror in their solar array to maximize the cumulative reflective power *in real time*. Traditional CSP plants have typically arranged mirror arrays in something of a "dead reckoning" arrangement in which calculations of mirror placement and the sun's position would determine where and how mirrors would reflect sunlight. This approach has been effective in allowing engineers to harvest high heats, but with the caveat of often losing a nontrivial amount of energy to "spillage" when some of the sun's energy does not reach the intended target.

Heliogen, on the other hand, uses advanced vision software to continuously obtain more accurate positioning information. With this data in hand, the AI calculates, on an ongoing basis, the exact alignment for every mirror in the array within a closed-loop tracking system. At the same time, the control software calculates micromovements for the mirrors to engage a more direct focus of sunlight onto a single point at the central tower's receiver. As a result, the heat transfer material can reach a higher "concentration ratio" and yield higher temperatures efficiently. In other words, with the help of AI and real-time measurements, more suns will effectively reach their intended target. Moreover, mirrors can also be less expensive given the added efficiency. An image of Heliogen's Lancaster, California facility is available on page 61 in the photo section right before this chapter.

The combination has allowed Heliogen to generate temperatures exceeding 1,000°C. That in itself is a game-changer. Industrial processes requiring temperatures that high (like cement manufacturing) have always been reserved for the power of fossil fuels. Now, engineers stand a solid chance of inserting a renewable, clean alternative to an otherwise carbon-heavy industry. That alone is a significant amount of energy and carbon emissions saved.

Better yet, the Heliogen team is attempting to do what has been commercially impossible to date: generate temperatures exceeding 1,500°C with concentrated solar energy. If they can manage to achieve that benchmark, we will all be in a new realm of clean energy. Because with temperatures exceeding that level, engineers will be capable of inducing on-the-spot thermochemical water-splitting—and potentially generating clean hydrogen at industrial scales.

How might Heliogen do it? The team takes advantage of a fascinating new material called ceria (CeO_2) that can be superheated to very high temperatures. When heated to the levels discussed here, the material will release a pure stream of oxygen. After cooling, ceria will draw this oxygen back—and water and CO_2 will break down into the components of hydrogen, carbon monoxide, and oxygen (which ceria then absorbs). The remainder of this transaction is just hydrogen and carbon monoxide, which engineers can also then recombine for synthetic fuel if they desire. But Heliogen understandably has their sights set on the hydrogen itself as a commercial product.

Heliogen is not the only company pushing for these developments. In April 2019, for example, "on-site" hydrogen production systems were demonstrated as proof in principle by researchers at the French Ecole Polytechnique Fédérale de Lausanne's Laboratory of Renewable Energy Science and Engineering (LRESE). Scientists were able to create a system estimated to run in excess of 30,000 hours (about 4 years) generating up to 1 kilogram of hydrogen per day with a concentrator the size of a large rooftop satellite dish—enough fuel for a hydrogen-powered car travelling 100 to 150 kilometers.[27] Naturally, operations would need to scale to meet consumer demands.

A number of other intriguing, potentially disruptive energy possibilities are still under investigation. Consider the possibility of harnessing electricity from plants directly, for example. A team of robotics engineers and biologists at the IIT-Istituto Italiano di Tecnologia in Pisa demonstrated that small amounts of electrical energy can actually be harnessed from the surface of plant leaves. As it turns out, certain leaf structures are able to convert mechanical pressure at the surface level into electrical energy. Through a process called contact electrification, electric charges on a leaf's surface are transmitted to inner plant tissue and subsequently throughout other parts of the plant. Pertinent here is that researchers were able to generate more than 150 volts just from a single leaf by harvesting this energy—enough to power 100 LED light bulbs simultaneously each time the leaf was touched.[28] By attaching artificial leaves to brush against others for additional mechanical force, researchers essentially mixed wind energy into the mix to amplify the effect.

Or how about generating energy from rainwater? Several years ago, Chinese scientists advanced the possibility of using solar panels to channel energy from rain. The team had placed a thin layer of graphene atop solar cells (we will get to graphene soon), simultaneously allowing for sunlight to pass through but also allowing the graphene to bind the free electrons within rainwater. The tiny amounts of positively charged ions in rain are then gathered through a process called the Lewis acid-base reaction. The resulting "pseudocapacitor" creates an energy difference between cell layers significant enough to generate small amounts of electricity.[29]

More recently, a new efficiency upgrade on harvesting energy from rainwater by researchers from Hong Kong's City University—on the scale of "thousands of times" better—has provided a huge improvement. The team developed a small generator system using a field effect transistor (FET)-like structure to connect electrodes with the incoming ion differential of rainwater. Their design utilizes a new material called PTFE with a "quasi-permanent" electric charge that can be connected to electrodes through a medium of raindrops. So a raindrop essentially hits the PTFE surface and subsequently bridges a chemical connection to attached electrodes, creating a full circuit. Incredibly, the team was able to produce up to 140 volts (enough to power about 100 small LED bulbs) from a single drop of water.[30] The device was not continuous, however—it required charging periods and did not, as yet, meet commercial standards to produce rooftop units. But research certainly continues.

Or more curiously yet: how about pulling electricity right out of the air? Ambient air often contains a noticeable amount of moisture. That moisture naturally makes the air slightly conductive, although electric charges are generally distributed evenly. Clever scientists at the University of Massachusetts have worked to gather ambient air vapors to produce renewable energy—"out of thin air," as they say—by connecting protein nanowires produced by the microbe *Geobacter sulfurreducens*. The bacteria grow these wires of protein to connect a current present in moisture to the device's attached electrodes. Their results, published in the respected journal *Nature*, described their intent to first power small wearable electronics before eventually scaling to phones and potentially one day larger applications for the device they have called Air-Gen.[31] And while we're unlikely to see those latter scaled-up stages for years, the prospect of developing yet another renewable energy source is certainly encouraging. The team was able to produce only 0.5 volts across a 70 micrometre-thick film—but noted that this effect can scale linearly as films are added to a collective in larger projects.

Concepts like smart clothes and electricity from plants or rainwater may just be novelties right now, but innovators are certainly innovating. Energy sources we never thought possible are becoming so. And regardless of whether we are generating our power from algae, AI-assisted mirror arrays superheating solar towers, or somehow squeezing fuel out of thin air, we now stand a greater chance than ever before of phasing out fossil fuels for sustainable alternatives.

CHAPTER 5

KEEPING THE LIGHTS ON

■

It is safe to say that the sun will be there for the next
4.5 billion years; therefore, we do not have an
energy crisis but an energy storage crisis.

—SURYA PRAKASH, USC (2019)

Pop quiz: what is the most widely used renewable energy technology in the world?

It might surprise you, but hydroelectric energy easily eclipses every other renewable with a global majority.[1] This is, in part, because harnessing the power of moving water to generate energy is as ancient as using the wind. Except that unlike the wind, large rivers almost never stop flowing.

Civilizations spanning vast distances of time and space have long used waterwheels by their riversides to transform the power of moving water into mechanical energy. They have used this energy to mill, grind, roll, and hammer various materials with partially submerged wheels attached to nearby levers, pulleys, and gears. As with windmills, our river-based energy designs took a milestone leap when waterwheel turbines were coupled with electric generators. Right now, in the year 2020, our hydroelectric engineers and scientists are still building upon many generations of innovation. And it shows: today's hydroelectric dams are remarkable engineering achievements.

We have only become more and more sophisticated with our hydroelectric designs over time. We have built larger dams. We have created side channels to store water we can use in the event of higher energy needs. We have pumped hydro to higher elevations to store even *more* water for times of greater need. We have even moved from simply utilizing the hydroelectric capabilities of surface-level rivers and have tapped subterra-nean water channels to harness moving water. Australia, for example, has a $3.1 billion project in the works to capitalize on water flowing between mountain reservoirs. The design will reportedly generate an additional ten percent of the nation's energy needs during peak hours while renewables may potentially be offline. The result will be something of a giant, under-ground battery waiting to kick in energy when needed.[2]

But while hydroelectric dams constitute the majority of renewables around the world, we still haven't been tapping the vast potential of mov-ing of water—perpetually coursing through our many rivers, pounding our coastlines at all hours, and swelling against the ocean's floors just off-shore. The force of all this moving water is untapped energy; something that wave, tidal, and river-based energy tech intends to harness. Because entirely aside from hydroelectric dam technology, the world's bodies of water present us with nonstop energy away from dams in the ocean and further upstream from dams in separate river systems. Capturing even just small portions of this power will augment a full-scale renewable en-ergy portfolio in crucial ways.

Given that river and tidal energy sources are so widespread, one might wonder why they are not a more significant part of our renewable energy conversations already. If we wanted to tap into a renewable energy source that never stops, does it not seem like tidal energy may have been a good place to start?

Part of our long delay on tidal energy has been technological. Designing machines to work in water can be notoriously difficult, particularly if that technology is fully submerged in water or if the water is full of salt. The pressure of waves continually bombarding tidal tech, the corrosivity of seawater, having our equipment relatively isolated from engineers who must suit up in specialized gear if they want to see it—all of these factors have limited extensive tidal energy campaigns. But the potential still ex-ists for never-ending waves to supply us with varying levels of energy to back up our grids 24 hours a day, 7 days a week.

Incredibly, the first patents for tidal tech date all the way back to the 1790s. Since then, patent offices have been flooded (sorry) with all kinds of new schematics ranging from "bobbers" and "flappers" to "water col-umns" and rolling "sea snake" tubes. All of them essentially harness the same kind of power in the force of water moving up and down or side to side, creating enough pressure to move internal mechanisms.

Some tidal energy technologies—variously referred to as marine energy converters (MECs), wave energy converters (WECs), or marine hydrokinetic energy in general—can operate on the water's surface, while others are placed well below. Techs below the surface can either be anchored and floating or fully anchored right against the seafloor. On the surface, waves rolling toward the shore will grow and collapse, pushing the blades of a turbine to generate electricity. Or, other surface-level devices will harness the power of breaking waves. Still others make use of swells, which are essentially longer wave patterns of energy that travel across the water's surface and move the water below.

At this time of writing, the world's largest tidal energy array is located near Scotland and features a series of seafloor turbines with blades up to 65 feet in diameter, churning out enough power for roughly 2,600 homes.[3] That's a tiny amount of power within the realm of renewables, but just as solar farms grow in time, the plan is to add more tidal turbines to power tens of thousands more. Massive new installations are currently forming around regions like the Hudson Strait in Canada, America's north coasts, the north coast of Australia, the Sea of Japan, Korea's Yellow Sea, and the northwest French coast along with the Strait of Gibraltar. At my last count, there were more than 80 new start-up companies in the US alone with plans to harness energy from waves. And as with the variation in patented designs, there now exists an assortment of companies trying out new techniques.

One simple design has tidal tech firmly anchored and lying flat against the sea floor. Researchers with the University of California recognized that muddy areas of the seafloor actually receive more pressure from incoming waves and may therefore generate more power.[4] Their design is something of a marine carpet, with surface material bending and flexing as tidal pressure moves overtop. As the rubber of the bottom moves up and down, it drives a piston pump to generate electricity which is then exported to the grid. The energy produced is not necessarily game-changing, but is stormproof, out of sight, and offers coastal communities the prospect of at least small amounts of continuous free energy.

Some companies and research groups have rolled out long buoy line designs with converters inside positioned to move against the pressure of incoming waves. Irish company Ocean Energy, for example, has developed a 125-foot-long buoy line for use in the Atlantic Ocean. The buoy line generates power as water pushes air through its turbine, resulting in spinning generators then converting the motion into electricity. And unlike many other tidal energy designs to follow, buoys are designed to move around—a lot. The line may move front to back or, with stronger waves, up and down. Multiple arrays of successive buoy lines can then essentially form wave farms anywhere one desires along coastlines.

Further inland, Scottish company Orbital Marine Power has championed a new river-based energy converter with other organizations in tow. Orbital's co-sponsored floating tidal energy commercialization (FloTEC) design, along with their own proprietary design, is essentially a floating boat hull that's anchored to the river bottom—with twin 20-meter turbines submerged but operating closer to surface level (which fortunately do not move with enough force to turn passing fish into sushi). Innovative designs like Oribtal's intend to catch faster-moving water near the surface while impacting marine ecosystems less near the bottom. (see page 62 of the prior photo section for an example image of Orbital's O2 design).

Spanish tidal energy giant Magallanes Renovables uses a similar design with anchored barge-mounted sets of turbines. The system boasts a much larger stationary centerpiece, however, allowing for very long turbine blades (and potentially more energy). British company Sustainable Marine Energy likewise uses a very similar concept—anchored tethers connecting a floating hull—but with multiple smaller sets of propellers underneath the design. If you picture a ship with two submerged wings and propellers dropping down on each wing to move with currents, you essentially have the concept.

Meanwhile, OceanBased Perpetual Energy in Miami has produced smaller and more modular sets of tidal energy arrays. The company's tech design basically resembles torpedo-shaped objects with propellers, spinning with the movement of water pressure—similar to miniature planes flying underwater. As the current forces an ebb and flow on rotor blades, full rotations can occur approximately 12 to 18 times per minute bidirectionally. And as with wind turbines, each one is connected to a gearbox transferring rotations to a generator.

OceanBased has plans for modular roll-outs within the Gulf Coast where southern US states meet the Gulf of Mexico. It should be an exciting project for Florida-based residents: investors pursuing the project estimate that the Gulf Stream can produce up to 45 terawatt hours per year of energy for Florida (roughly 2 to 3 nuclear power plants' worth of energy).[5] Tidal energy estimations are often undervalued: they can be built upon piece by piece, scaled to a meaningful degree, are often stormproof, and cumulatively provide energy 24/7. In what has been called the "Gulf Stream gold rush," wave energy companies now clamor for the chance to reserve a patch of seabed the way offshore wind companies desire placements for turbines.

Isolated rivers do not even need to be large to offer adjacent communities and homeowners a source of energy. Smaller, more "micro-hydro" designs again offer smaller amounts of energy, with energy just waiting to be harnessed.

Belgian micro-hydro company Turbulent offers its own spin on aquatic power: small diversions built into existing rivers able to capture essentially never-ending side channels of water energy.[6] Picture a small, brief side-channel built onto an existing river. Water flows partially into that side channel, through a turbine system, then back into the main river itself. This design can be implemented repeatedly along the same stretch of river, if desired. Turbulent's design offers a decentralized, rural-area solution for nonstop renewable energy in localized rivers and canals. The average result, they say, is enough to power approximately 60 homes.[7] Moreover, Turbulent claims that the installation only takes about a week and is fish-friendly.

Or consider Japanese tetrapod tidal energy converters. Beneath various locations along Japan's coastline, structures called tetrapods function as "wave breakers" to help ease the force of incoming waves. It makes sense, then, to outfit wave-breakers with energy converters. Researchers at the Okinawa Institute of Science and Technology Graduate University (OIST) developed equipment capable of doing precisely this. Their tetrapods more closely resemble flower-spinners anchored to the ocean floor, utilizing the force of crashing waves to move turbines while the structure itself helps to buffer the shoreline from erosion.[8] The equipment again utilizes softer and more bendable blades to avoid harming marine life or birds that may attempt to land on the structures.

Seattle-based Oscilla Power uses another wave energy tech appealing enough to land on Project Drawdown's spotlight campaign for coming energy attractions. Their design is considered solid state—made without external moving parts and potentially more long-lasting as a result. First, picture a platform sitting flat on the water's surface with a long tether connecting it to the ocean floor next to a coastline. Underneath the bobbing platform is a counterweight—and as the surface platform bobs around, the counterweight pulls back against the movement of waves.

This recombinant jostling about moves the energy-converting components inside through compression and decompression.

Other new innovations increasing the viability for tidal tech include more sophisticated instruments used to locate ideal spots for turbines. After all, faster-moving water with more abundance will help generate more power and offer a quicker return-on-investment for financers. Research teams have advanced the capabilities of computer modeling programs like SWAN (Simulating WAves Nearshore) to index ideal locations within 5 km of shorelines, water depths of no more than 22 m deep and more.

There are two crucial reasons why I believe tidal and wave energy are simply waiting for the right combination of engineering. One, because water is nearly one thousand times denser than air, water-based turbines are theoretically far more efficient than wind turbines.[9] And two, new data science can allow for tidal energy *planning*, as internet-of-things (IoT) cloud connectivity with weather systems can allow tidal tech to forecast incoming waves up to five days in advance.[10]

Ultimately, the future of tidal energy simply means improved engineering, many additional roll-outs, not squandering useful rivers and waterways—and potentially largescale installations harnessing multiple styles of wave energy simultaneously. In terms of the latter, one compelling new project is the Cooperative Research Center, led by Associate Professor Iren Penesis of the University of Tasmania's Marine Renewable Energy Research Group. The group is pioneering new "living laboratory" prototypes to develop an offshore space in which renewable tidal energy can be combined with aquaculture. They even have plans for underwater drones to take care of many of the lab's operations.[11]

That's what the future looks like. Combined use of offshore wind turbines, solar inputs, an assortment of wave energy converters, and even a feature for hydrogen-production using seawater—all while developing research on these advancements and cultivating aquaculture at the same time. Doing so much more with so much less.

GEOTHERMAL INNOVATIONS

As we explore the potential of our tides to provide us energy that's admittedly smaller but still perpetual, we now come to another source of essentially nonstop renewable energy with similar appeal. Geothermal heat emanating from within the Earth is unlikely to stop any time soon. And as with tidal energy, geothermal energy is chronically under-discussed given its potential applications—and chronically under-funded in its research as a result.

Geothermal energy can help us in four ways. First, thermal energy can provide us with electricity directly through a variety of thermoelectric tech designs. Second, thermal energy can supply us with heat, reducing our need to spend electricity warming our air. Third, thermal energy can actually be stored to supply us electricity when other renewable sources run out. Fourth and perhaps more importantly, thermal energy has intriguingly been used to facilitate community-level "thermal grids" to strengthen electric grids. More on that last part soon—let's focus on scientific and technological developments for the first three factors first.

First: as a source of "baseload" energy, geothermal is always on. Plants converting thermal energy to electricity can operate 24 hours per day, 7 days per week, year after year. Moreover, geothermal can be a steady and predictable energy source.

The world has come a long way since the first geothermal electric plant was constructed in Italy more than 100 years ago. At that point, the energy output was barely enough to power about five light bulbs. America's current geothermal capacity is now more along the lines of 3.7 GW—enough to power 407 million LEDs.[12]

Geothermal sadly constitutes a comparatively very low percentage of the world's renewable energy at the moment. In fact, major geothermal installations are actually hard to find. In a country like the US, for example, major geothermal sites are grouped largely around just a few sites in California, Utah, and Nevada. With further R&D initiatives, however, the Department of Energy (DoE) has estimated that geothermal energy can potentially account for more than 10 percent of the entire country's electricity demand.[13] Incredibly, DoE analysts also believe that with enhanced geothermal systems (EGS), potentially more than 100 GW might be deployed.[14] For a sense of scale, 1 GW is equal to about 3.125 million photovoltaic panels.[15] One might also calculate the energy and carbon savings from building retrofits to take advantage of geothermal and add to that total.

For reference, enhanced geothermal systems are wells in which engineers drill and then subsequently inject water from the surface. Normally,

we might hope that the Earth's heat alone can power turbines at surface-level. But in areas where that is not possible, enhanced systems cycle
water down to cause subsurface fractures in the rocks below to release
more heat. While this fracturing method is one borrowed from Big Oil
(fracking is the term used), these fractures are only intended to increase
permeability in much smaller spaces. So industry experts believe it is not
quite as dangerous as the hydraulic fracturing performed during oil-
oriented fracking.

Regardless, after water has been cycled down below, it is pumped back to
the surface with heat to spin turbines. If engineers are cycling to far lower
depths, the temperatures they get back may come closer to "supercritical"
and yield greater energy (though engineers have yet to reach officially
supercritical depths). The real benefit of EGS, aside from tapping more
geothermal energy, is that geothermal systems can be placed in more locations, rather than relying on reservoirs of hot water below.

Countries just on the outskirts of the European Union like Iceland and
Turkey have invested heavily in geothermal energy. Turkey in particular
has roughly doubled its geothermal capacity between 2016 and 2018—
reaching an energy output capable of powering several million homes.[16]
In fact, Turkey's current geothermal capacity at this time of writing is
more than the entire 28 countries of the EU combined.[17]

Elsewhere, governments have been more cautious in pursuing geothermal applications. In Europe, geothermal energy still remains a fringe option amidst a renewables market dominated by wind and solar. It accounts
for only 0.3 percent of the EU's primary energy production.[18] This is partly
because hot springs underground are not always easily accessible—meaning steep government approval processes to begin this work have ham-
pered development. There are also a few dangers to consider in geothermal
operations. The initial drilling process, for example, can cause effects
similar to those of fracking. A South Korean geothermal plant was actually
responsible for a magnitude 5.4 earthquake back in 2017.[19] As a result,
geothermal engineers have become wary of unique ground compositions,
the potential for faults or fractures, unusual reservoir depths or ground-
water issues, and more.

Cost is another constraint for geothermal applications. It costs money
to dig deep into the ground and provide an air-exchange system to one's

home that can then be regulated by sensors and feedback systems. New developments for geothermal, then, have managed to cut into these costs with innovative designs.

Berkeley-based Fervo Energy, backed by large sums from the Breakthrough Energy Ventures (BEV) associated with Bill Gates and other investment heavyweights, uses a non-vertical drilling design inspired by the oil industry to lower costs. The Fervo team uses a mixed-mechanism stimulation method popularized by the shale industry, supposedly improving geothermal drilling productivity up to four times. They, like many other up-and-coming geothermal start-ups, use distributed fiber optic sensing equipment to help monitor geothermal flow rates for easier installation.

Alternatively, one of the more promising companies offering advanced geothermal tech is Swedish start-up Climeon. Climeon uses a closed loop system in which water is contained within a 2x2x2 container. The container is exposed to geothermal heat that boils the water in one chamber. The resulting steam from that boiling water moves a turbine inside, which generates electricity. After steam passes through the turbine post-electricity, it reaches a cold chamber in which the water vapors cool back into a liquid. The cooled liquid water is then shuttled back into the original compartment to maintain a continual output of electricity. This essentially never-ending process of generating electricity can, for every individual 2x2x2 cube module, power about 150 kilowatts. That is roughly enough to power about 250,000 mobile phones.[20]

Climeon's design can actually be utilized in a number of different settings. Waste heat, for example, can power the process if it is hot enough. At least one Finnish cruise line uses the tech, piping waste heat from their ships' operations to a Climeon unit taking in small amounts of sea water for the liquid feeding cycle. The company claims that the subsequent steam power, aside from helping to propel the ship, also reduces CO_2 emissions by about 4,000 tonnes per year.[21]

Geothermal heat not only provides us with a source of electricity to harvest, but a source of heat to spare us the electricity in generating our own heat—specifically if used to heat our homes. Consider Alphabet/Google's Project Dandelion, which sprung from the still relatively secretive "X" division where techs with big dreams are given a long leash to go after so-called moonshots. Separate X divisions have gone after various renewable energy plans, though only some of which have led to commercial results. And in terms of geothermal heat, Dandelion is one such project.

The research team behind Dandelion found that by testing out a number of different drilling options (some of which again were perfected by fossil fuel interests), they managed to substantially lower the cost of tapping geothermal energy on a by-house basis. Dandelion's crew tried a variety of options: modified, high-powered jackhammers, liquid nitrogen ground-freezing tech—at one point, the team even tried highly pressurized water jets to "obliterate the ground."[22] Eventually, the team discovered that a fast and slender drill tech that vibrates can cause top-level soil to act more like a liquid, making it easier to cut through. As a result, the team is able to produce a hole just a few inches wide to the desired depth in mere hours, rather than several days. And as a side bonus of completing the work in a few days with less damage, less land needs to be remediated thereafter.

Consider that geothermal energy can also be combined with the new technology for harvesting waste heat discussed earlier—and the fact that geothermal energy can be present in lower-level heat at depths we normally inhabit. This is the concept behind Swiss-based company Enerdrape placing heat-absorbing panels in underground parking lots and subway tunnels. The company cites an ability to recycle heat more efficiently throughout buildings if that heat is directed through the pads toward ventilation systems. The subsequent heat channeled to buildings above can replace up to 250 kWh of energy otherwise used for heating through panels that cost only 150€ (about $163). Enerdrape will reportedly have its first pilot project ready by the end of 2020 (though again, that may be delayed now).[23]

Another new innovation for geothermal energy is the prospect of swapping out traditional drilling tech with high-frequency microwave beam generators—to "melt and vaporize" rocks, as Seattle-based AltaRock has proposed. Rather than using a mechanical drill, AltaRock uses extremely high frequency millimeter waves produced by a gyrotron (essentially a specialized microwave beam generator) to open up holes in large chunks of rock. The team is still experimenting at lab scales as I write these words, but the innovation may ultimately allow engineers to reach deeper depths more quickly. Incredibly, on AltaRock's successful grant papers sitting with ARPA-E (the Advanced Research Projects Agency-Energy organization within the US government), the team has estimated that just 0.1 percent of the planet's heat could provide humanity with our total energy needs for 2 million years.[24] For now, the world waits with bated breath for new test results from the team's trials at the Oak Ridge National Laboratory in Tennessee.

Or how about mixing geothermal energy with hypersonic propulsion research? You may have heard in the news that military organizations around the world are developing "hypersonic missiles." These missiles travel much faster than traditional weapons, making them far more difficult to intercept or shoot down. But at least one research team in the US is working with hypersonic propulsion to explore how fired projectiles may ease the task of drilling underground. Seattle-based Hypersciences is reportedly working on a drill system that fires concrete projectiles at more than a mile per second (that's greater than Mach 5) into rock formations in front of a drill. The projectile hits rocks with enough velocity to essentially vaporize on impact, while any leftover remnants are mulched by the subsequent drill. Hypersciences claims their design can drill to lower depths up to 10 times faster than existing models, creating greater geothermal potential at more locations.[25] Moreover, after the drilled well is complete, Hypersciences then uses pipes containing silicone oil (rather than the water that other enhanced geothermal systems often use) to route heat back to the surface. Once at the surface, the team uses thermoelectric generators rather than turbines to create electricity.

As with all other renewable energy techs, incredible new advancements in nanoscience and materials sciences will yield better equipment for us to use. More durability. Better conductors. Lighter materials. Stronger parts. The same applies to geothermal's ability to provide viable renewable energy, even if some of the prospects admittedly dip at least partially into the sci-fi realm.

For example, have you heard of *graphene*? Graphene is a material straight out of science fiction only now becoming more widely used in advanced engineering. It is essentially a smaller, shaved-down version of graphite, a material used on the tip of pencils. In smaller form, graphene is shaped like a honeycomb of all carbon atoms so thin it was heralded as the world's first two-dimensional object. It is almost transparent and nearly weightless while still being stronger than steel and is one of the best conductors in the world. Three-dimensional objects made with graphene can be approximately ten times stronger than steel—making composites of the material excellent candidates for exterior finishing applications.[26]

Graphene stands to revolutionize materials in ways we still can't even imagine. But one of its most amazing properties is also one of its least-reported: graphene is an excellent conductor of heat. So good, in fact, that you are able to melt an ice cube with the heat from your fingers simply by holding a graphene sheet against the ice.

Now, just keep graphene in mind while we get back to geothermal energy. A few miles below our feet is Earth's nearly unlimited heat. We can tap into that heat and use it for energy, if we can manage to transfer it closer to the surface sustainably. Copper instruments we have used for these purposes would simply melt at certain depths due to the extreme heat. As would many other materials we might use to conduct the heat below. Graphene, on the other hand, has a higher melting point and thermal conductance. As a result, a state-of-the-art engineering team might be able to drill deep enough into the Earth to tap its vast amounts of heat—and bring it to the surface safely and continuously with a material like graphene.

The CEO of Five Hour Energy Drink, Manoj Bhargava, has popularized the concept of using graphene to collect geothermal energy through a documentary called *Billions in Change*.[27] I recommend checking it out, if you haven't already. Bhargava details using his wealth to fund a workable prototype for graphene-based geothermal applications that is still reportedly ongoing today. At the moment, though, even our best engineering teams are still working on drilling deep enough to safely deploy energy equipment. So this concept, sadly, will be on the shelf for a while.

ADDRESSING OUR STORAGE PROBLEM: BETTER BATTERIES AND P2X

We have now explored some continuous, 24/7 sources of energy to augment a grid increasingly reliant on intermittent renewables. But it still won't be enough to fully transition from fossil fuels if a perfect storm of energy loss happens. In the event the sun isn't shining, the wind isn't blowing, and we lack enough wind or solar inputs from somewhere else on an increasingly interconnected international grid system—and the available power from sources like tidal, geothermal, and nuclear cannot fill the gap . . . we'll be in trouble.

Part of filling our energy gap can be powerful new lithium-ion battery backup systems. These systems have often not been feasible at a societal scale to date, though it is true that battery systems have declined in price over the years and are projected to decline further, though not to the degree that making battery backups for large grids is an easy decision. So while typical largescale battery systems can help, they typically cannot yet sustain most entire grid networks.

Critical assessments of the potential behind battery backup systems often fail to mention several things. They usually do not discuss the increasing number of alternatives to traditional batteries. Nor do they cover many power-to-x (P2X) grid backup options, an increasingly versatile suite of new options for backing up our energy. Nor, for that matter, do renewable energy critics often discuss how powerful new grid upgrades will change our valuations of exactly how much energy we need and when we'll need it.

Some advocates point out that massive battery installations really can augment an entire grid, if only they are developed. Consider the Hornsdale storage facility in Australia—the result of a wager between billionaires Mike Cannon-Brookes and Elon Musk. After plenty of banter about the power of Tesla batteries, Musk bet Cannon-Brookes that Tesla technicians could have a grid-level battery system up and running in Australia in one hundred days. And in fact, Tesla's engineers did just that, delivering a grid-level backup system in well under the 100-day deadline. What's more, it's already been used successfully.

The Hornsdale installation is essentially a giant battery contributing stored energy during periods of high grid load. The technology has provided that energy with "speed and laser precision in response to system events," according to local electric operators.[28] Moreover, over the span of its creation until this time of writing, Hornsdale's storage system has saved operators an estimated $40 million in so-called "grid stabilization costs" (safety backup measures undertaken to avoid blackouts).

Sadly, while the battery market will deliver sizeable dividends to a successful new design, the market also has an incredibly high failure rate for new ventures. Start-ups consistently form and fall in the battery market, always seemingly offering the next big substitute for lithium-ion batteries. Even right now, several compelling new start-ups are campaigning on vastly improved energy efficiency rates while using cheaper and more Earth-abundant materials. MIT spinoff Ambri focuses on using liquid metals, for example, outperforming conventional lithium at grid scales.[29] But unlike other industries, I urge battery innovation enthusiasts to keep expectations in check until commercial products are ready (or at least nearly ready) to hit the market. Research into sodium-ion flow batteries is also palpably close—this iteration has progressed towards mass-commercialization and sulfur-ion batteries are already acting as grid-cover in the Middle East.

Even solid-state battery tech, perhaps the most famously wide-ranging and disruptive battery upgrade of them all, can be a tough sell for investors. Solid-state batteries—another technology always "just a few years away"—replace the liquid electrolyte you will find in typical batteries with a solid interior. These batteries have been touted to pack in up to 50 percent more energy over traditional batteries and also charge faster; both potential game-changers for traditional battery systems. So while the technology has been tough to develop, the promise of immense value for one lucky company or patent-holder keeps plenty of labs out there working on it. One secretive new spin-off company from Stanford University scientists called QuantumScape has generated plenty of buzz after their own solid-state battery provided a 466 mile drive for VW's E-Golf vehicle—a marked improvement over its typical 186 mile single-charge distance. QuantumScape's official production target will supposedly ramp up for commercial applications in 2025, according to a press release.[30] But time will tell, of course.

Crucially, labs and engineers are increasingly beginning to question exactly what a "battery" needs to be. After all, if we're just talking about stored energy we can use when necessary, who cares what form that takes? Does it always need to be a chamber of chemical electrolytes sitting in a capsule?

Consider that we do not often consider hydroelectric dams doubling as batteries. Normally, we only think about how flowing water hits turbines and generates electricity. And yet, most hydroelectric dams also cleverly include a stopped side channel of water waiting to be released during times of increased energy demand. Many will also pump additional water to a higher elevation during times of plenty, again just holding onto it for later release if needed. This "potential energy" of stored water functions like a battery because it's essentially just energy waiting to be used.

Energy storage (read: battery) innovations have therefore taken a number of fascinating new forms. We are branching out from the chemicals of traditional batteries and hydroelectric backup sources. We are combining different mediums of storage to create a series of backstops for the times in which we really need spare energy. The concept of "power-to-x" has therefore built upon the concept of using the excess energy we *do* have during times of renewable energy plenty (sun shining, wind blowing)—and using that excess energy to do additional work. Namely, work that will result in excess energy for us when our renewable sources are not around. And as it turns out, we can power a variety of different operations to store potential energy and keep it waiting for use. The operations, in this case, represent the "x" in power-to-x.

If we were to use excess renewable energy during times of plenty to produce hydrogen gas, for example, that would be called power-to-gas (P2G). This can essentially be a roundabout form of energy storage. Because if excess energy can power the production of hydrogen, that gas can then be combined with carbon dioxide at a later time through methanation to produce synthetic natural gas. That's fuel we can use at any point thereafter, stored away. If the CO_2 used in that process is captured from ambient air, the gas can burn carbon neutral while providing us with an advanced form of energy storage and smoothing out more of our renewable energy intermittency. Research firm Wood Mackenzie has projected that by 2025, another 3,205 megawatts of renewably powered electrolyzers will help to produce so-called "green hydrogen" around the world (up from just 253 over the past two decades).[31]

Excess renewable energy can do more than just create gas and pump water to higher elevations for gravity to drop it, though. Renewable energy can also lift a variety of other weights to higher and higher elevations. New and innovative designs are now capitalizing on storing energy this way. What if, rather than water released to flow over turbines, we released objects held at great heights to slowly pull on gears they're attached to? Those gears will spin just like turbines, producing energy in much the same way—without the need for a river.

Picture a series of cranes hoisting concrete blocks, powered by renewable energy and simply waiting to release those blocks as needed. This is the setup of Swiss company Energy Vault, a new grid-level energy storage system that augments other storage mediums. The stored energy of a mass that has been raised can be cashed in when the object is driven downward by gravity and activates a motor. More weight lifted means more stored energy; higher elevations mean increasing the potential energy when releasing the weight. Hence peak energy hours can actually produce more stored energy. Advanced AI sensors can then activate energy production in under three seconds to supply grid-level energy if needed.[32]

There are multiple benefits in using a "gravity battery" to store energy. For one, there are no toxic chemicals used. In fact, bricks used can be made of waste materials. Second, the potential energy does not degrade in the same way a thermal or chemical medium may. Third, the engineering is incredibly simple. And fourth, the design can be implemented essentially anywhere.

Clever engineers have also used the power of gravity to energize batteries in other ways. EV construction vehicle engineers, for example, have utilized the "regenerative braking" style charging of hybrid cars released in recent years. Regenerative braking is basically a type of hybrid vehicle model that charges your battery not only as you drive using gasoline—but every time the brakes are pressed. Getting into the engineering specs of how pressing the brakes recharges your battery would require a lengthier explanation than space will allow here. The bottom line is that the same mechanism from pressing a brake can also be used when gravity pushes a vehicle down a hill. A vehicle with a heavy load (topped up at the top of the hill) will apply more pressure to a regenerative braking system as it moves down a hill. Hence, Swiss engineers have designed an electric dump truck able to recharge its battery upon descent to power its ascent back to the top.

Now, pumped hydro can also take on more forms than just diverted side river channels and raised-elevation tanks. Newer innovations on pumped water as an energy storage medium have started to gain traction. Houston-based Quidnet, another BEV-backed energy company, applies the concept of storing water but without the need for geographically-limiting rivers or dams. Instead, Quidnet's system uses excess electricity to pump water underground into shale rock. The company uses either new wells dug specifically for the purpose or old, abandoned oil-and-gas wells. Water essentially fills up tiny cracks in these rocks, creating pressure that will force the water back out. This process compresses the shale almost like a spring. Pressure is then released as needed, resulting in the water gushing out to waiting turbines.

New research has actually revealed that Quidnet might be on to something. Seasonal pumped hydropower storage (SPHS) officially received the academic greenlight from a study concluding the tech can be an affordable and sustainable option for storing grid-level energy over long periods of time. The study, published in *Nature Communications*, found that significant potential exists around the world for storing energy through pumped water channels.[33]

Meanwhile, investors are also patiently waiting for a prototype to emanate from similar projects circulating on a global scale. For example, the World Society of Sustainable Energy Technologies has filed patent papers for an application called EarthPumpStore. The project will use abandoned mines as compression-storage sites rather like Quidnet and has already analyzed applications in the US, UK, and China. And why wouldn't they? The University of Nottingham has estimated that China alone has about 150,000 unused mines—which, if they were fitted with this technology, could store roughly 25 times the energy China requires each day.[34]

Similarly, a partnership of international tech giant Siemens and Spanish renewable tech giant Gamesa has focused on compressing hot air into volcanic rock. Pumped air is heated with spare grid energy and the volcanic rock formations hold the temperature at roughly 750°C/1382°F for about a week. When the facility, located in Hamburg-Altenwerder, is required, it uses a steam turbine to re-electrify that stored energy and restores it to the grid.

Technology designed to harness the potential energy of pumped air is often called compressed air energy storage (CAES). And as with all other technologies discussed here, inventors and innovators have been experimenting with new techniques to improve it. One challenge with CAES is that the heat generated by compressed air is often subsequently lost during the storage process—with fluctuations of air reserves then impacting how much air pressure is available to produce energy.

To circumvent these problems, Toronto-based Hydrostor created an "advanced compressed air energy storage" (A-CAES) system. Their version uses a mix of tailor-made chambers designed to strengthen pressure and release mechanisms, captures the heat that builds up with air compression, and then recycles it, as well as adding in a liquid water component to help even out the pressure supply. As a renewable energy storage device, excess energy runs Hydrostor's compressors and then pressure is simply released when a grid backup is required. A key bonus being that the approach may be much cheaper and simpler than working out the schematics for pumped water storage systems. As I write these words, Hydrostor is working on securing new installation projects ranging from 300 MW in power to 500 MW. While the technology is yet to be proven as a reliable grid backup for those levels of support, many analysts see no reason why they (or a similar company) could not.

Next to gravity and pressure-based systems, we also have the energy storage power of heat—which certainly isn't a new concept. Nineteenth-century engineers were already storing steam to drive turbine generators. But these systems have always been very inefficient. There are always steam-release issues and heat loss. New developments, fortunately, appear to store and release heat much more efficiently.

One of the more fascinating thermal energy storage innovations I've seen has to be the "Malta" project spinning out from Alphabet's X division. Project Malta focused on storing heat within molten salts designed to release thermal energy over longer periods of time. An electrothermal system first takes electricity from a renewable source and a heat pump turns it into thermal energy. That thermal energy is then stored in the molten salt medium. In times of need, a heat engine converts the energy back into electricity for the grid.

Malta's molten salt tech can therefore also be located anywhere without the need for hydroelectric rivers or deep chambers. Moreover, the design dynamic takes advantage of the cooling differential of cold air on low-temp liquids for more energy. And the components for molten salt systems are all readily available, so it's fairly cheap. Most of the tech's components included in each system are just easily extracted salt, steel tanks, and cooling liquids. Now spun-out as Malta Inc., the company even boasts that each salt tank can be re-charged "many thousands of times," offering buyers a shelf life of roughly 40 years.

It's worth noting that the heat storage capabilities of molten salts are now becoming a selling point for advanced nuclear reactors attempting to offer multiple roles. Some molten salt reactor companies, including Terrestrial Energy in Canada and Moltex in the UK, have marketed their ability to not only provide energy while the grid is in operation—but then having stored thermal energy thereafter within a reactor's molten salt fuel.

New consumer products also exist for augmenting grid irregularities with something of a personal energy bank within one's home or business. The Tesla Powerwall, for example, already allows for stored solar energy to be used during off-peak hours. But other "solar+storage" devices will increasingly exist in thermal energy storage and other, newer forms. Suddenly, Manoj Bhargava's pitch for an electric bicycle one might pedal for their workout at home while feeding energy into their house supply seems less outlandish than it did a few years ago.

Best bets for future energy storage devices therefore seem to include stackable weights, compressed air, pumped hydro, and one final tech yet to be discussed at length: flow batteries. Flow batteries are like large traditional batteries using cheaper materials to shuttle chemical energy where it needs to be. Flow batteries have stuttered in development over the years, relying on volatile markets for key components like vanadium. Newer flow batteries have delved into iron or sulphur with promising results, although work continues. Scientists have found that flow batteries could power hundreds of homes for hours on end in the event of a grid outage, should the technology be developed further.

A related but more fringe technology is the "entropy battery" under development at Stanford. As noted, channeling rivers to pour heavy water onto turbines has been a time-tested human favorite. But newer options have begun to show promise for alternative forms of renewable energy, made possible by the fact that water is a good medium for shuttling electrically charged chemicals around. Newer renewable energy options discussed earlier—like harvesting the power of rainwater and ambient air moisture—are showing particular promise.

Scientists have even found a renewable energy source within the salinity differences in the places where freshwater rivers meet saltwater bodies. Clever engineers at Stanford have therefore noted that potential energy exists within the aqueous interchange between freshwater rivers and the saltwater bodies they flow into. The "salinity difference" threshold constitutes a static energy source, able to be captured with entropic flow batteries. The team estimated that if the world's rivers were harnessed for salinity differentials, their technology could provide roughly 2 TW of energy (or about 13 percent of the world's energy consumption). Time, again, will tell.[35]

The "mixing entropy battery" is first filled with freshwater. It is charged with a small electric current and drained, then filled with seawater (technically not renewable if we're priming the process, but I digress). Because the saltwater contains about 60 to 100 times more ions than the freshwater previously in the battery, the voltage increases between two electrodes. The battery then provides an output of more energy than was required to discharge it initially. With the help of new advancements in nanotech, the battery is able to generate a current from salinity differences essentially anywhere a river is flowing into the ocean. The mouth of the Amazon River, for example, where the world's largest drainage basin flows into the Atlantic Ocean, could be an energy-producing hotspot for this technology.

SMART GRIDS, MICROGRIDS, AND THE IMPENDING ELECTRON ECONOMY

We have established that our energy sources are becoming much more "distributed" over time. Solar inputs, wind inputs, tidal inputs, and more are all feeding into our electrical system. We are therefore moving further away from centralized power providers toward more decentralized sources. This process is driving other crucial innovations: a smarter grid to handle all our distributed sources and smaller microgrids that can be more reliant without major providers.

Normally, most of us don't think about where our electricity comes from. We just connect the plugs from our appliances into a wall socket and let the electricity companies do their thing. But that ambivalence about where our energy comes from is about to end.

The way our grid system typically works is that electricity is harnessed from an energy source—be it from coal, natural gas, nuclear, or renewables. Once electricity is generated, it flows through an energy provider and enters the regional electric grid. That energy all tends to "look" the same to energy purveyors; electrons are electrons, after all. One might think of renewable energy inputs being added into this mix rather like pouring a cup of water into a river. Those renewably produced electrons pour into an otherwise bustling river of energy flowing toward a middle-man site—the energy company purchasing the electricity in bulk contracts to ensure its river is nice and full—before the flow arrives to your wall socket.

Today's current grids were not designed to handle a multitude of energy inputs, however. And current grids also just communicate in one direction: from producer to consumer. Developing smart grid technology will improve on this by allowing two-way, machine-to-machine (M2M) communication to take place between producer and consumer—as well as allowing for two-way power transfers. As a result, average consumers with access to renewable energy inputs will increasingly become able to commercialize their own energy to the grid or even just local users. But what really puts the "smart" in smart grid is that these connections will be far more automated and handled by artificial intelligence.

Artificial intelligence makes up a particularly diverse scientific field. Within the realm of AI, we might find machine learning or even more advanced deep learning. Machine learning simply allows computers to learn from and interpret sets of data. Advanced coding can allow software to learn rules from its interpretations of the data it receives, with rules piling on top of those rules. In the real world, machine-learning applications allow massive, facility-wide monitoring systems to detect abnormalities in typical system function. Say, by receiving a ping from a sensor leak and alerting human operators. Or, even better yet: the smallest disturbance in the data obtained from sensors can trigger an automated response essentially as fast as electrons can travel. With constant streams of data inputs shaping machine-learning patterns as they develop, newer systems are even capable of producing predictive models for future patterns. Simulations produced by machine learning are then helpful for engineers and contingency planners looking for systemic weaknesses in their designs.

In terms of a smarter grid system, artificial intelligence will be much faster than human operators at detecting variable inputs and outputs. It will also be faster at assessing the information obtained from consumers' energy use to provide predicting models over time. And it will be more efficient at communicating with user-side devices to adjust and sync power supplies (and sometimes, demands). All of which contributes to what energy experts call "load flexibility"—the ability to forecast impending energy needs and be flexible about the timing of energy supply as a result. Grid AI will also use historical records and weather patterns to form predictive flexibilities. Sensors connected through IoT will communicate even small disturbances in a renewable source's ability to generate power, letting a grid network know well in advance of pending issues.

Let's consider an intelligent grid using the data of typical peak times and tasked with ensuring that energy is available as needed. Standard, daily charging and powering-up times could be automated to run at different times—if, say, water heaters are set to run at particular times overnight, a smarter grid connecting a whole set of these devices on a community level could taper them out in flexible terms rather than charging all at a potential peak time and at the whim of individual consumers. Meaning that a smarter grid may opt to power up your water heater at 4 a.m. one night, or possibly 2 a.m. the next to smooth out energy peaks and valleys. Functions like EV charging will undoubtedly be synchronized in similar ways.

More device communication will also allow smarter grids to create price-of-energy fluctuations to match demand. As in, tailored to cost less over night when less people are using energy—encouraging consumer behavior that becomes more predictable for an AI-driven smart grid. Energy demand will be more predictable if many users have their settings adjusted to charge things during lower energy costs, for example, providing energy producers more insight about scheduled demand.

Of course, smarter grids require several developments. They need responsive appliances and home sensor systems, but they also need advanced metering going to those homes and high-voltage power lines with enhanced long-distance transmission capabilities along with the necessary sensors. So, while smart grids are basically inevitable, they're still waiting on critical infrastructure upgrades.

I was recently able to speak with Distinguished Professor of Energy at the University of California, Berkeley, and former Science Envoy for the State Department Daniel Kammen about the future of grid energy tech. He explained to me that while an internet-of-things interface will increasingly expand how we relate with tech in our everyday lives, an "internet-of-energy-things" will increasingly expand how our tech relates to the grid.

Potential energy generators will become increasingly diverse, he told me, with power providers capable of making energy sales through renewable inputs they own. The transformative new energy features of homes, cars, buildings, sidewalks, and roads will all involve new inputs. All of that energy can be pooled and coupled and potentially sold if produced in excess of personal or community-level demand. The integration of an immense number of renewable energy inputs will more readily make "baseload green energy" the norm, according to Kammen.

Second, Kammen elaborated that consumers will increasingly become "prosumers"—actively selling energy either back to the grid itself or to other local users. Ideally that will inspire far more renewable energy input installations. After all, why be the only homeowner not saving money on your monthly bills *and* making money sharing your excess energy with others?

New smart grid tech companies are betting on this equation. A company like Australia-, Singapore-, and San Francisco-based GreenSync full-on embraces energy fragmentation through their exchange cloud platform, aggregating a multitude of energy assets. The company believes that future smart grids will connect literally millions of energy resources per grid into an automated and seamless energy flow, revolving around electricity retailers and prosumers big and small.[36]

Better yet, grids increasingly capable of producing these outcomes can further inspire states, provinces, or entire nations to increase the practice of exporting energy. A larger, more national (or even international) grid system can further smooth out the intermittency of renewable sources—because even if the sun isn't shining anywhere near your grid, perhaps it is in another state or country with excess energy to sell.

On a larger level, cross-border energy commodification could empower any nation to be an energy exporter. Just as individual homes and businesses have an incentive to invest in renewables, the economic incentive for countries to empower their communities with energy-storing capabilities should increase over time. The Dogger Bank Wind Farm turbine array being built just off the coast of Yorkshire, for example, will start generating electricity in 2023 and provide power for 4.5 million homes each year while linking to grid systems for Norway, the Netherlands, Germany, Denmark, and Belgium.

On a much smaller level, even highly localized "nanogrids"—encompassing residential homes or individual commercial buildings—will help to reduce energy expenditures and save owners money. The benefits we will gain from using smarter and sometimes smaller independent grid systems like peak energy shaving, time-of-use synchronization to only run equipment during optimal price times, and more can all be scaled down to help smaller systems.

THE NUCLEAR ISSUE

And now we come to the elephant in the energy room: new advancements in nuclear power. Nuclear energy stands to play a pivotal role in our transition away from fossil fuels—but is also vulnerable to a few (very) unpredictable variables.

Supporters of nuclear energy are often enduringly devoted to furthering its cause as a clean, safe, 24/7 source of limitless energy that we need to develop immediately. Not only because we can, but because nuclear can succeed where renewables sometimes fail. Passionate nuclear supporter and co-founder of Breakthrough Institute Michael Shellenberger rightly argues, for example, that an average large-scale solar farm takes up hundreds of times more space than a nuclear power plant—among many other criticisms.[37] To say nothing of the added mining and refining of parts, developing infrastructure, etcetera.

On the other hand, nuclear advocates tend to gloss over the advancements of renewable power and energy storage. And in many ways, renewables can be a lot less complicated than they are often portrayed by nuclear supporters.

Nuclear energy really is a wild card. If nuclear sites were operated impeccably, the risk of any danger is quite low and the transition from fossil fuels comparatively very easy. The issue is that nuclear can seem like a great idea—until it isn't. Conventional plants with human operators are not always run impeccably. Accidents can happen. Safer designs are still in development and relatively untested. Revolutionary new tech designs are still a question mark, despite large flows of venture cash. Perennial game-changer nuclear fusion has broken our hearts many, many times before.

First: definitions. Fission is the type of nuclear energy many of us are familiar with. We often see these kinds of plants in movies and shows depicted with two large silo-like structures venting gas. Nuclear fission involves splitting heavier atoms into lighter ones, transforming energy in the process. The processes of enriching materials like uranium and using them in fission power plants, however, comes with caveats that generally make us uncomfortable—for example, the possibility of a catastrophic malfunction, harmful gases and nuclear waste, weaponized material, and more.

Nuclear fusion, on the other hand, involves fusing atoms together and harvesting the subsequent energy. Fusion is what powers the sun and other stars and, unlike fission, does not create nuclear waste or materials that terrorists or rogue states can weaponize. Certainly no Fukushimas or Chernobyls.

The primary challenge with nuclear fusion is that it's a very difficult process to orchestrate. Fusion requires incredible heats topping many millions of degrees Celsius. And then, heated materials must be kept near to the plasma contained within a controlled chamber. These processes have always required more energy than they produce, making fusion "net negative." However, incredible new advancements in superconducting materials (magnets especially) and powerful new machine-learning algorithms have furthered fusion research considerably. A game-changing development, then, would be achieving nuclear fusion at lower temperatures and becoming "net positive" in producing more energy than what was used.

Several exciting new companies are working on it. Commonwealth Fusion Systems (CFS), for example, has raised more than $100 million to work on producing net-positive energy outcomes. Their hope is to have a live prototype utilizing next-gen superconductors by 2021 and a full plant to show off by 2025. If it works, this will be an incredible accomplishment—and perhaps a force moving the needle on decarbonizing our energy at rapid speed. An MIT spin-off, CFS enjoys Bill Gates-level funding and mental giants behind the wheel. Canadian company General Fusion is also taking advantage of superconducting magnets and new upgrades in artificial intelligence to further its nuclear fusion research. They are also eyeing a potential mid-2020s net-positive reactor. An image of General Fusion's unique design can also be found on page 63 in the photo section prior to Chapter 4.

TAE, previously known as Tri Alpha Energy, raised an astounding $600 million in venture funding to expand its fusion research. Tri Alpha's labs are backed by groups like Paul Allen's Vulcan Capital and the Rockefeller family's Venrock, among others.

Reactors like those of General Fusion and Commonwealth Fusion Systems are called tokamaks, typically donut-shaped and reliant upon high-powered magnets to control circular streams of plasma. TAE, on the other hand, will essentially pulse 100 million watts of electricity into a cloud of plasma—30 million degrees, for reference—shooting protons into the mix. The collision force of this reaction fuses single-proton hydrogen into helium, producing immense amounts of energy. Which, despite the sound of it, would be rather safe: a leak of energy will simply degrade the plasma into a solid that is, as scientists on the team have cited, sized at less than a grain of salt.[38]

The largest fusion research project currently underway is the International Thermonuclear Experimental Reactor (ITER) under construction in France. The project is a collaboration among nuclear specialists from 35 different countries working to help revolutionize the world's energy supply, if they can. The ITER is a tokamak reactor aiming to generate 500 MW of fusion power from 50 MW of input heating power by the year 2025. Although, generating more power than what is supplied is not the same as perpetuating that reaction and doing so continuously at larger scales.

Many scientific and engineering commentators seem to believe that a commercial-scale fusion reactor operating at a power plant scale may still be a decade away or longer. Nevertheless, a new kind of nuclear race is moving at full speed for the research teams and private companies hoping to bring the world's first successful fusion reactor online. As of right now, Commonwealth is predicting their latest reactor will exceed the energy "breakeven rate" by 2025. UK-based First Light Fusion is aiming for even sooner in 2024. TAE is hoping for 2025. Naturally, many of these companies could very well fail and go bust at any point before these dates—but if even one of them succeeds, humanity in general will benefit immensely.

Thorium fuel is another promising nuclear avenue. Thorium is another chemical element which, like uranium, can produce nuclear energy for us. But unlike uranium, thorium is incredibly plentiful. It is easily mined and is about three times more common. Thorium has a longer half-life. The nuclear waste produced by thorium requires storage of only a few hundred years compared to the roughly ten thousand of uranium. Thorium waste can also be combined with pre-existing nuclear waste to be re-used after it's been "spent." And ultimately, enough thorium could reasonably be obtained to power reactors for thousands of years with cheaper operating costs. While thorium fuel cycles are not entirely "proliferation-resistant," vulnerabilities thereof are fewer and further in between.

The use of molten salts to replace water as a coolant has also been another innovative move for advanced nuclear reactors. Because molten salts can absorb more heat safely and reactors can more closely reach atmospheric pressures, there are fewer safety concerns (and hence, fewer risk mitigation costs). TerraPower, for example, was cofounded by Bill Gates and is currently working on a new reactor variant utilizing molten chloride over water as a coolant. Their hope is to develop a prototype by 2030. Moltex Energy in the UK is also working toward a 2030 deadline, estimating that their molten salt reactor will only cost about one-eighth of a current reactor for the same energy output.[39] Terrestrial Energy in Canada is furthering their Integral Molten Salt Reactor (IMSR) design, which offers both grid backup capabilities and a transfer of energy-intensive heat for sites up to five miles away. Florida-based ThorCon reportedly plans to begin testing a molten-salt-fueled fission reactor by 2023–2024 (and was recently supplied a huge grant from the US Department of Energy to do so).

Nuclear can absolutely be an ace in the hole if teams continue to develop it. A cursory glance at the International Energy Agency's tracker for nuclear energy, however, has it listed as "not on track" and mentions of fusion are nowhere to be found.[40] A new release from Project Drawdown titled "Drawdown Review 2020" described nuclear energy thusly: *Nuclear power is slow, expensive, risky, and creates radioactive waste, but it has the potential to avoid emissions from fossil fuel electricity.*[41] That about sums up this author's love/hate position on nuclear energy, but I'd love to see continued funding for any promising new technology—nuclear included.

So, just how possible is it to transition away from fossil fuels? Traditional renewables like solar and wind have always offered us a realistic off-ramp to wean ourselves partially from carbon-heavy fuels. But can we augment those intermittent energy sources with smaller but *continuous* sources to truly break free?

If nations were to pursue more hydroelectric energy, build out more tidal arrays, deploy advanced geothermal equipment, develop new energy storage mediums, and establish better electric grids to manage supplies and demands where possible, then I believe we can realistically say that yes, it can certainly be done. Furthering some or all to a greater degree pushes us closer toward a goal of energy independence.

On the other hand, if all these amazing new developments for renewable energy have gotten us closer to fully integrating clean energy—have there been any new advancements for fossil fuel plants in becoming less polluting and more green? In fact, there have been several: new developments in improving fossil fuel efficiency, pollution reduction, and carbon emissions reductions have transpired across the board. Combined cycle plants often utilize waste heat to drive a turbine and off-set some of the energy that would otherwise be wasted. Clustering exhaust capture systems can potentially channel the emissions of multiple industrial sources into more economic scales of filtration, making some heavy-emitting energy plants less wasteful. Though perhaps no fossil fuel development has been more alluring than the "Allam Cycle" energy plants now being developed and tested by North Carolina-based company NET Power.

Named after lead inventor Rodney Allam, the Allam Cycle is an energy plant design that swaps the burning of a fuel source like natural gas in open air for burning that fuel in a chamber of pure oxygen, with CO_2 as the carrier gas. Normally, fuel is mixed with air and then direct-fired for combustion. But by using "oxyfuel" instead, NET Power can essentially eliminate some major air pollution by-products, like nitrogen oxides, of traditional plants. An Allam Cycle plant will then produce nearly pure CO_2 instead, which can then be piped elsewhere (like a storage or utilization site). In this case, carbon dioxide is captured as part of the combustion cycle and not through a separate, energy-intensive process.[42]

Additionally, piped-in CO_2 is pressurized and combusted at supercritical temperatures (where distinct liquid and gas phases do not exist and CO_2 is referred to as sCO_2). This process then uses pressurized carbon dioxide to move a fluid turbine and generate energy in the process—before the waste fluid (water and CO_2) is ultimately separated for re-use. Multiple outlets for recycling waste heat help drive this process at various points along the way. The remaining CO_2 not used (then supposedly at a purity level of about 90 percent) can be safely piped away.

In terms of efficiency, newer combined-cycle power plants can often run at about 62 percent efficiency while NET Power will run at roughly 59 percent efficiency—but without the emissions and air pollution.[43] At the moment, only a small test plant exists to demonstrate this process as proof in principle. But a commercial-scale plant is being projected for about 2023 in western Texas. Time will tell if fossil fuels can become carbon neutral during power generation—my guess is that they cannot, and we'd better keep rolling with renewables. But we should always continue to improve where we can and hope for the best.

CHAPTER 6

INDUSTRY WITHOUT EMISSIONS

■

The American way of life is not up for negotiation. Period.
—U.S. PRESIDENT GEORGE H. W. BUSH (1992)

I magine a world in which streetlights no longer switch on at sundown. Or imagine that streetlights no longer generate artificial light for our streets at all. Imagine, instead, that our unused streetlights have been wrapped with genetically hybridized vines designed to emit a soft, green glow of bioluminescent light to illuminate our nights entirely without the use of electricity.

Picture a sustainable nightlife without the massive energy drain that artificial light requires. Imagine, in this futuristic green society, that we have genetically augmented these hybridized vines with the same code of DNA found in the deep-sea organisms that glow to survive. Imagine that we have engineered these intrepid new vines to emit their low-key radiance the moment a visual light cue of dusk arrives—wrapping and overtaking lampposts that we once spent up to 6 percent of all total energy in a country like the US.[1] Think of reducing the world's carbon emissions by about 5 percent and, of course, the additional benefits of having plant life surround us in urban areas.

Sound like science fiction? The idea of glowing plants replacing our city lights seems to have been popularized through a synthetic biology project crowdfunded back in 2013. The funding campaign was massively successful at the time, easily surpassing the tens of thousands of dollars its DIY biohackers had originally benchmarked. The "Glowing Plant Project" certainly had high hopes when proposed, but was later scrapped due to the genetic difficulties in generating sufficient amounts of light. As it turns out, using genetic editing to splice bioluminescent DNA into organisms can help in some ways but not others. Prototype plants did not provide nearly the level of brightness we see in animals deep in the inky black of our world's lower oceans. As of mid-2017, our dream of all-natural, *Avatar*-esque plants bringing a whole new meaning to the phrase *night life* was off the table.

When crowd-funded, do-it-yourself biologists bowed out, however, the establishment team of mental giants at MIT tagged in. They brought with them a conceptual framework for embedding light-producing nano-particles directly into plant cells—eliciting a chain reaction of biochemistry to generate at least enough green light to read by. The dream was alive again. The group at MIT led by chemical engineer Michael Strano had access to newly developed wavelength-specific nanoparticles, capable of generating various degrees of light in response to the light (or lack thereof) reaching plant cells. Which means, potentially, that vine-wrapped lampposts (or just flower beds) that "turn on" at dusk may yet be a reality. Self-growing, self-repairing nightlights thriving by sunlight and water alone while drawing in CO_2 from the surrounding air, ready to light up our nights for free. Imagine it again, if you like.

Of course, glow-in-the-dark plant lamps are still a long way off; but any small step we can take toward reducing our energy expenditure at a societal level, while trying to keep in place parts of our lives that we enjoy, makes the transition away from fossil fuels that much more doable. Progress toward replacing something energy-intensive and carbon-heavy in our lives with something low-energy and carbon-light carries us closer to that goal. Especially if we manage to move from something carbon-heavy to carbon negative.

Currently, at this time of writing, Strano and his team are still actively spearheading research into "plant nanobionics." Strano's lab has actually experimented with a number of different research avenues aside from the now-famous "Light Emitting Plant" project. They have found that plants can do more than merely emit soft green light. By embedding various plant species with different kinds of nanoparticles, the team has enabled living plants to manifest unique, non-native functions, like monitoring for and notifying plant growers of drought conditions in the soil below. And, while I suspect this may not have been the team's original focus—one batch of signature spinach is capable of detecting explosives and communicating that information to nearby smartphones.[2]

It may sound incredible, but biological organisms are now helping us do even more than detect bombs and a loss of soil moisture. For example, would you believe that we are designing viruses capable of self-assembling battery components, almost entirely eliminating the need for large-scale factories full of toxic chemicals in the process? Or that engineered algae and cyanobacteria are capable of producing anything from plastic to phone screens—all while simply thriving on sunlight, water, and CO_2?

With a suite of new technological tools at their disposal, scientists are introducing us to a new class of organisms that can clean up wastewater (and produce energy while doing so), make plastics, create consumer products, and generate useful pharmaceutical chemicals, fragrances, and even food ingredients. With their help, we stand a solid chance at slashing the energy and emissions costs of many sectors in our industry profile. Incredible new advancements in genomics, nanotechnology, synthetic biology, and the emerging new field of molecular engineering are making these feats possible.

Industrial manufacturing carries an immense carbon footprint providing us the products we enjoy. After transportation and our electricity infrastructure, industrial manufacturing takes up another 22 percent of total American carbon emissions.[3] Re-thinking how we produce the apparent necessities of twenty-first century life will therefore involve some seriously out-of-the-box thinking. It will mean re-thinking all the chemicals and materials going into our products, the global agriculture industry producing our foods and assembling the consumer goods we enjoy, and more.

As usual, scientists attempting to improve upon the wasteful aspects of our lifestyles are looking to nature for clues. After all, many scientists are now increasingly asking themselves, how is it that nature can produce amazing materials all on its own without factories and electricity? How is it that such incredible materials—hard shells and feathers and exquisite natural chemistry—can be produced in nature without all the industrial run-off and smokestacks? Without the supply chains and shipping channels? How is it that evolutionary blueprints actually go about making versatile materials like spider silk and abalone shells?

Evolutionary biology has had billions of years to produce novel solutions to complex environmental problems. As a result, evolutionary designs are often well-worn and battle-tested for functionality. And scientists are increasingly paying attention to the details. So the question now, is: why haven't we been copying the kid in class who knows all the answers?

RE-THINKING INDUSTRY: SYNBIONANOTECHFTW

If you're a newcomer to evolutionary biology, fear not. The crash course on why we see such incredible biological designs throughout the animal kingdom essentially involves the tandem forces of chance and error playing themselves out over very long periods of time. The passage of genetic information from generation to generation is necessary for the continuation of life, but is never achieved with complete fidelity—copying errors and chance genetic mutations continually occur as parent organisms pass genetic material to offspring. This may sound bad at face value, but the engine driving successful evolutionary adaptation is that some of these randomly occurring mutations turn out to be very beneficial. Beneficial mutations then carry forward to subsequent generations and life builds upon small, incremental steps toward the wonderful complexity we see today.

The vast majority of genetic mutations are not at all advantageous. This is partly why evolutionary history is so breathtaking; we are often seeing the chance winners of a genetic lottery in the environments we admire. Indeed, anywhere we see animals thriving in apparent complementarity with local environments, we are seeing an incredible series of very gradual successes.

Throughout life's journey from the Earth's first cells to modern humans, cellular factories have produced everything from the simple chemicals sustaining tiny microbes to the ornate and protective shells on marine life or the silk of spiders. Some of the successful operations that cellular factories have learned to master are things we now wish *we* could do: harnessing energy from different sources, storing and releasing that energy on command, transforming energy into food, metabolizing other unique chemicals in the process—and bigger chemical operations like the catalysis driving complex chemistry, oxidation, synthesizing advanced materials, and more.

As MIT's visionary biological engineer Angela Belcher has famously wondered: *If abalone can make all the shells they need over millions of years without emitting toxic by-products . . . why can't humans make everything we need without polluting our environment?*[4]

More and more often, we are.

Consider bioplastics. Even if you haven't heard of them, you have probably used them without knowing it. The "plant-based bottles" you've potentially seen at the grocery store are often bioplastic at work—made when the sugars taken from plants and crops (like sugarcane or corn) are fed to microorganisms living in a vat somewhere. The organisms are engineered to ferment those sugars into materials that can later be skimmed from the solution's surface as plastic resin. That resin is then processed into bottle shapes and other designs. Much of the energy and infrastructure normally spent processing petroleum into plastic? Gone.

Other times, scientists will use algae to metabolize specific chemical products instead. In these cases, algae or cyanobacteria have simply been bioengineered to process their food in ways that result in the production of particular chemicals thereafter, like the photofuel process described earlier. More energy and infrastructure spent producing those chemicals the traditional way? Gone.

With even slight alterations of a microbe or alga's genetic code, scientists have produced alterations in the materials they can produce. This fine-tuning of genetic formation leading to a fine-tuning of chemical products has given biotech engineers an incredible array of new powers. A range of genetic combinations and subsequent material results now exist in databases worldwide. And after years of research, some teams now have whole (often privatized) databanks of genetic combinations for producing particular materials. Moreover, some teams are even bolstering the production of biomaterials we never thought possible with the help of AI.

How sophisticated have we become? One lab/company specializing in state-of-the-art biotechnology is Boston-based Ginkgo Bioworks. The company recently became more famous (and filthy rich) by partnering with pharmaceutical giant Bayer to produce sustainable ingredients. The company's CEO Jason Kelly describes customers as approaching his team with requests along the lines of, "program me a GMO to do x"—and engineering a biological solution through designer microbes.

Want a fragrance without the industrial footprint? Kelly describes having induced a yeast cell to produce rose oil for perfume manufacturers by splicing new segments of DNA into the yeast's genome. Or perhaps you might want something entirely new to nature—like a lily-scented fern for novel indoor plants. That's only a sticker price of $10 million away from development, if you're interested.[5] In Kelly's words:

> *Cells are programmable similar to computers because they run on digital code in the form of DNA . . . We believe Ginkgo has the best compiler and debugger for writing genetic code and we use it to program cells for customers in a range of industries.*[6]

Kelly argues that industries as separate as clothing, electronics, food, and pharmaceuticals are really just biotech industries that don't know it yet. I am inclined to agree: using engineered biology to produce chemically identical plastic at a lower price with fewer emissions seems like a no-brainer. Or, if you did want that lily-scented fern (or whatever fantastical new hybrid you believe will be a bestseller), companies like Ginkgo are really your best shot.

Biological equipment offers a number of additional benefits for materials manufacturers. After all, organisms are self-assembling, self-replicating, and self-repairing. Materials can be tailored just as well or sometimes even better than through traditional manufacturing. As a result, the genetic codes being stockpiled by companies like Ginkgo now exist in a growing "bioeconomy"—offering unique organisms for sale, pertinent for a variety of industrial projects.

Berkeley-based Caribou Biosciences is another growing biotech giant using CRISPR technology to engineer yeast strains for industrial fermentation. They, too, create chemical factories for bio-based fragrances and flavors, or even industrial cleaning and transportation products.[7] A company like Boulder-based DMC Biotechnologies uses microbial fermentation to make a broad variety of bio-based products, including flavors, fragrances, nutraceuticals, and more. At this time of writing, the company reportedly has more than 800 unique microbial strains capable of catalyzing all kinds of new products. All the excess energy and industrial infrastructure we might otherwise expend producing those products? Gone.

Seattle-based Arzeda even utilizes artificial intelligence to churn out new *protein combinations* to use in designer organisms. Machine-learning applications can map out the physics of molecular systems in nature and sift through potential biochemical combinations in advance, advising bioengineers on how best to build the new designer chemicals they need. Some combinations have actually never been seen before in nature. The team's software is investigating *all* possible stable combinations. It's mind-blowing, when you really think about it: protein folding sequences and their amino acids are incredibly complex. An AI capable of spitting out viable and pertinent protein chain designs is very impressive AI.

What can Arzeda do with such capabilities? Among other things, the team is currently using microbes to help produce materials for phone casings and scratch-proof screens. In partnership with the US Department of Energy, Arzeda is investigating new applications for tulipalin, a molecule normally found within tulips. Plastic-like biomaterials augmented with tulipalin are actually stronger and more durable than average phone casings and screens if mixed in the right ratios. So the team analyzed the metabolic pathway in tulips responsible for making tulipalin and spliced that component into the industrial microbes they had back at the lab. Engineered microbes then produce that same molecule inside fermentation vats, without the rest of the tulips. What we'll ultimately have, if the team is successful, are scratch-proof, semi-glass screens on our smartphones—brewed in a lab without the massive industrial footprint.

Or consider San Francisco-based Checkerspot, whose slogan is "Designing new materials by accessing nature's molecular palette." Like Arzeda, Checkerspot uses advanced computational algorithms to plumb the depths of what is biologically possible. Incredibly, the team found that they could engineer a strain of microalgae to produce a new signature form of triglyceride plant oil.

Normally, we have been limited to cooking and manufacturing options from just a few plant oils existing in sunflowers, corn, soybeans, rapeseed (read: canola), and palm oil. A greater selection of plant oils could offer new and more sustainable (or even potentially healthier) options for a variety of applications. The Checkerspot team found that engineered microalgae could produce triglycerides normally harvested from fields of crop materials—without the fields, the lag-time, or the physical harvesting. In other words, Checkerspot's alga strain can produce a high-performance material without all the energy, infrastructure, and carbon emission requirements of the traditional means of production.

Checkerspot's outdoor brand WNDR Alpine launched their first official product in July 2019—a line of eco-friendly skis, largely made by algae. The lab has since partnered with Gore (of the Gore-Tex line of apparel), presumably to produce more outdoor-capable fabric and gear materials, again, largely from algae.

The promise for biomanufacturing is revolutionary. Not only because labs like these can replace traditional industrial infrastructure, and therefore reduce a significant amount of carbon emissions; but also because the product they sell is just as good, but cheaper. And research teams are just now getting started exploring new and potentially game-changing molecular combinations.

It's worth noting, before we discuss how scientists intend to feed the post-carbon world, that most material and food processing has been limited to just 14 fatty acids over time—but hundreds exist in nature. What potential combinations may be just on the horizon for bioengineers exploring new chemical compositions? You may recognize "fatty acids" in cuisine as making up the saturated, trans, polyunsaturated, and monounsaturated fat contents in foods. Fatty acids serve several purposes within the body in providing energy for the muscles and the heart, and storing energy elsewhere. We have been limited to using or consuming the fatty acids provided to us by nature. However, an ability to safely tailor the chemical composition of new oils could provide what some scientists are calling "tailored health benefits."[8] Of course, research into designer chemical compositions like these remains in infancy for now.

Research from an independent team at the University of Bonn furthered Checkerspot's approach with cyanobacteria, rather than engineered algae. The team was able to produce plant oils with lab-grown cyanobacteria—again allowing plant oil distributors to ignore land-based restrictions.[9]

Getting back into industrial manufacturing: Checkerspot's labs are not the only teams growing and brewing clothing material in labs. Rather than using fossil fuel oil-based nylon for jacket threads, Japanese start-up Spiber has used brewed proteins similar to spider silk that can be fermented in large vats. The outcome is a material that can be skimmed from the surface and refined into a threading with similar form and function to traditional silk. A company like The North Face Japan (under Goldwin) can use these specialty designer threads in their jackets, again with a substantially lower carbon cost than traditional methods.

Even more incredible is that when Spiber scientists found that recreating an identical version of the spider silk protein did not work—the material soaked up water, usually a hindrance for jackets—they redesigned the protein molecule to create a hydrophobic version of spider silk. Just like that. Changes in chemical sequence could result in materials that are stronger, more heat resistant, stretchier, and so on.

Materials with a massive environmental cost like cotton can also potentially be replaced with biotech. Products relying on animals raised on ranches or plantations can be replaced with more sustainable alternatives produced by biotech. Some of them already are. Clothing threads can even be designed to biodegrade in the right conditions, sparing us some landfill space (and no, a biodegradable bikini will not simply melt off of you at the beach).

New Jersey-based Modern Meadow, another bio-fabrication lab/company, grows plant-based leather in-house using engineered yeast. The company makes Zoa, a material with the same form and function as leather, but without any use of animals. They begin by engineering yeast DNA to create proteins they can then have arranged into the necessary material formations. Modern Meadow can even engineer their plant-based leather to tweak small variables in weight, toughness, or thinness. This brings us a little closer to freeing ourselves from the animal-product necessities in these industries—and their associated carbon footprint.

Peoria-based Natural Fiber Welding has also popularized Mirum, a plant-based leather faithful in form and feel, scrupulously upscaled from agricultural by-products. The material is cleansed and processed from a composite of different chemical blends, bonded together through a series of controlled chemistry steps. The result is a strong leather essentially just like "normal leather" with a light texture but no plastics or animals involved. That, again, helps to reduce the amount of energy or emissions required for producing a material many people desire.

There are many other biotech labs and companies working to replace energy-intensive and carbon-heavy manufacturing processes. Discussing them all here was not possible, but no product seems to be off-limits when bioengineers set their sights on sustainable alternatives.

WHEN BIOLOGY MAKES THE MATERIALS AND MANUFACTURES THE PRODUCTS

If you think that brewing our clothes, designing new houseplants, or growing ski plastics with algae are exciting projects—then it's time to buckle your seatbelt.

While the last section of this chapter covered a few ways in which microbes are helping us to produce the materials we need to eventually create products, MIT biological engineer Angela Belcher's lab is designing organisms to actively *manufacture* those products.

Consider Dr. Belcher's question: How is it that abalone can take materials simply found on the sea floor and produce a variety of useful shells? How are those shells actually manufactured on a biochemical level—and merely from biological components following a set of genetic instructions? Would it be possible to program our own organisms to manufacture similar objects if we simply provided them with the materials and the genetic instructions to carry out the work?

If you have already seen Dr. Belcher's TEDxTalk lecture on the subject, you'll know that it *is* possible. And that the Belcher lab utilizes an unusual workhorse to accomplish real manufacturing duties: viruses.

Viruses are quite different from most other organisms in that they do not possess some of life's crucial components. Unlike the functional units we find in cells, viruses are simply a protein chain encasing genetic instructions in the form of RNA or DNA. As such, they lack the capacity to thrive on their own or reproduce without a host body. They are, simply, a set of instructions for carrying out specific actions.

Because of their comparative simplicity, viruses can be easier to work with and manipulate on a genetic level than essentially any other kind of organism. As a result, various viruses have been rendered harmless for lab work and manipulated for experimentation. Nowadays, we are testing beneficial viruses to carry genetic information to particular cells—like transporting immunotherapies into the body or targeting cancer cells with genetic kill codes. So-called "phage therapy" uses therapeutic bacterial viruses to treat pathogenic infections and may be our answer to antibiotic-resistant strains of bacteria.

In Belcher's lab work, she noticed that some viral variants could pack materials densely together. Rather like the abalone, she thought, making the necessary nanostructures of calcium and carbonate for their shells. If those materials were nanoparticles of various metals instead, the instructions could be workably similar.

Belcher's lab then took to the painstaking work of assessing billions of viral strains to isolate the unique properties for a successful adaptation. Engineer countless strains, isolate successful variants, repeat. The end goal was to discover and engineer enough viral components to pack together not just calcium and carbonate—but the metallic constituents of batteries.

Generation after generation of engineered viruses were strategically selected for their dedication to a single task of moving select materials around. And although that may not sound like much, when millions of engineered viruses are gathered together and completing the same task, the cumulative effort brings a significant amount of directed manufacturing. Belcher's viruses would eventually assemble the anodes, cathodes, and casings of batteries, even if they didn't really realize what they were doing.

I cannot do Belcher's lab justice here, so I recommend checking out their work. For example, they may have plans to one day cultivate viral strains capable of manufacturing autobody parts and even *living batteries*.[10] Other researchers also believe electroactive microbes could one day produce electricity. Living colonies of electron-producers self-repairing and living to discharge electricity.[11] I highly encourage looking over the citations for some additional research.

In short, Belcher's lab was able to produce viral strains capable of self-assembling battery components. The process avoided the high temperatures of normal manufacturing, the organic solvents, the toxic chemicals and, most importantly, the excessive carbon footprint. The more our molecular engineers scope out natural patterns of nanoscale materials placements, the closer we are to having sustainable alternatives for more products. Belcher's lab now uses viruses to work with more than 150 different materials, including virus-produced solar cells (though we can now make cells cheaply enough that the lab did not commercialize their process). The question then becomes: if we take on this process for battery casings, what else can we manufacture in a lab? What else can we shrink the energy expenditure and carbon footprint for?

Incredibly, scientists at the University of California, Riverside, have altered a virus to arrange gold atoms into "spheroids" measuring just a few nanometers in diameter.[12] This should allow production capabilities for items like electronic components to be cheaper, faster, and easier than before. Senior author of the paper, Elaine Haberer from the nearby Rosemary Bourns College of Engineering, summed up the work well:

> *Nature has been assembling complex, highly organized nanostructures for millennia with precision and specificity far superior to the most advanced technological approaches . . . By understanding and harnessing these capabilities, this extraordinary nanoscale precision can be used to tailor and build highly advanced materials with previously unattainable performance.*[13]

In fact, we may soon be able to direct the production and/or manufacture of materials rather like a conductor conducts a symphony. Remember the Strano lab and embedded nanoparticles in plants? Consider that scientists may potentially combine the use of embedded nanoparticles with the engineering of designer organisms to create new possibilities. As it turns out, an external stimulus can influence the embedded nanoparticles—to impact internal cellular function in predictable ways.

Realistically, a variety of microorganisms other than plants can support embedded nanoparticles. Scientists at the University of Colorado Boulder demonstrated in mid-2019 that they could selectively activate particular enzymes with the help of embedded nanoparticles in microbial cells.[14] In other words, researchers have taken steps toward creating not just living factories, but living factories we can then influence thereafter. The point of which would be to further tailor unique, non-native functionality into self-assembling, self-repairing materials.

In terms of the science behind this research: quantum dots (tiny semi-conductor particles) are injected into cells passively. They are designed to attach and self-assemble to desired enzymes—which scientists can then activate on command in response to specific wavelengths of light. The quantum dots essentially operate like spark plugs, activating particular enzymes to accomplish cellular work in response to whichever wavelength of light is used. In the case of this study, a red wavelength could amp up cellular production of converting CO_2 and producing bioplastic. A green wavelength could amp up production of ammonia (a fertilizer). The team's microbes would simply lie dormant in water without photoactivation. They release the results of their chemical transactions to the surface, where the materials can be skimmed off and refined. The researchers at CU Boulder therefore essentially created light-powered nano-organisms capable of consuming CO_2 while producing sustainable bioplastics and other materials in the process.

Scientists have also demonstrated that it is possible to assemble artificial biostructures with DNA—and that nanowires constructed from the material can conduct electricity within a nanoscale electric circuit. One experiment from Columbia University in New York, for example, fabricated DNA nanowires to conduct electricity in place of valuable metals and silicon. The use of synthetic biostructures may one day help to create a "renewable electronics" industry in which nanotech and synbio combine to build biological wires (or "BioWires") for small, self-assembling electronic equipment.

Biotech aside, there are plenty of other reasons to be hopeful about our capabilities in reducing the energy expenditures and carbon footprint of industry. Sure, synbio and molecular engineering will clearly help reduce significant portions of both. But materials sciences are advancing in many other ways in the meantime. Consider that today's aluminum cans are produced from 75 percent recycled content. Simply from advancements in materials sciences alone, we have reduced the greenhouse gas emissions profile for aluminum cans by an astounding 90 percent over time.[15]

Other new technological developments will help reduce waste and decarbonize industries in indirect ways. Major tire manufacturer Michelin plans on releasing its puncture-proof, airless tire around the year 2024. Airless tires will require fewer raw materials and result in fewer new tires needing to be made for replacements.[16] Developments like these are so innumerable that cataloguing them all here would certainly help make the case that transitioning to a post-carbon world is increasingly possible—but also run this book well over its readable limit.

One study from 2014, for example, found that with more of the world using a tech like 3D printing, global manufacturing costs could go down by about $170–593 billion US.[17] Industry energy use could be reduced significantly thereafter—and CO_2 emissions could be reduced by up to 525 megatons (Mt) by 2025. Changes like these—materials advancements, substitutions, potential for biological design—will force us to recalculate how much energy we will need and how bad our emissions will be going forward.

Even difficult-to-decarbonize industries like steel and cement are bound for incredible new sustainability changes. Remember Heliogen and their clever use of AI in orchestrating sophisticated mirror arrays to generate extreme solar heat? I mentioned that such a process could be a game-changer for carbon-heavy industries. And it's true: a new development like advanced heat-absorbing material coupled with AI-driven CSP heat stands to revolutionize carbon-heavy industries.

An industry like cement manufacturing, for example, takes a heavy toll on the environment. Project Drawdown cites cement as contributing to about 5 to 6 percent of *all global carbon emissions*. As a result, producing an alternative form of cement ranks #36 on Drawdown's Top 100 list of climate change solutions.[18] Finding a greener solution can shrink industrial output of emissions at large significantly.

What often makes cement so wasteful is that part of its process is roasting (typically) limestone and clay at high heats, necessitating a considerable amount of energy. As noted back in the section on concentrated solar power for renewable energies—extreme heats are a primary part of this process. The extreme heat of a CSP plant can now theoretically be used to produce cement more sustainably, reducing potentially up to 40 percent of cement's emissions.[19]

A company like Heliogen may or may not succeed—after all, nine out of every ten start-ups reportedly fail. But pioneering new tech has been developed and the blueprints are out there—solar-powered extreme heat to decarbonize the cement and steel industries is just one of those success stories. Brewing clothes and electronics components in a lab is another. Some researchers are even growing products through engineered mycelia (mushroom) networks that can produce anything from clothing threads to housing bricks.[20] Creating designer viruses to shuttle around tiny metal components into useful shapes is yet another looming success story. We, as a species at large, are increasingly replacing uniquely wasteful aspects of our post-industrial lifestyles. From the device you might be reading this on to the clothes in your closet or the petroleum-derived cosmetics in your washroom—forging a bio-basis for any of these products can help each step of the way. Little by little, we're moving toward a more livable world.

CHAPTER 7

FEEDING THE
POST-CARBON WORLD

■

Unlike the cow, we get better at making meat
every single day.

—PATRICK BROWN, CEO, IMPOSSIBLE FOODS (2017)

The industries we need to move our civilization forward tend to be the areas where most of our emissions are. Transportation, electricity, and manufacturing are the top three sectors contributing to carbon emissions. It should be no surprise, then, that industrial agriculture comes in at a close fourth. How we grow, package, ship, and consume our food is a significant source of not only American carbon emissions, but global emissions as well.

The EPA estimates that agriculture's role in American carbon emissions clocks in as high as 9 percent of the total—which, given the sheer size of America, equates nearly with the cumulative emissions of Britain.[1] Globally, the Food and Agriculture Organization of the United Nations estimates that roughly 14.5 percent of all anthropogenic greenhouse gas (GHG) emissions emanate from agriculture and livestock.[2]

Fourteen-and-a-half percent is a colossal number. How and where we get our food is clearly a largescale, systemic climate issue—to the point that eight of the top twenty solutions in Project Drawdown are food-related. Drawdown's researchers, like the scientific community at large, recognize that the world's population is still rising rapidly. And that means not only more food, but also more energy required for growing, packaging, and shipping it all—along with a subsequent rise in emissions. Some scientists even argue that future populations will require more nutrients and more food as the world heats up.[3]

Meanwhile, our global agricultural system is also not very sustainable for other reasons. We are consistently exhausting the nutrients in our soils, thereby reducing the nutritional value of what we eat, and we have rampant pesticides that we cannot always eliminate from our foods. We have synthetically boosted our crop yields to meet the demands of growing populations but have done so with unsustainable materials. We use manufactured nitrogen, which requires a large amount of energy (and carbon emissions) to produce. We suffer the subsequent "run-off" of excess nitrogen from our fields seeping into rivers and larger bodies of water, choking up organisms, and creating marine dead zones.

Worse yet, we will need to contend with the added challenges of climate change in how we grow and deliver food going forward. In August 2019, the UN's Intergovernmental Panel on Climate Change released a lengthy report on food insecurity in the decades ahead as land is degraded and climates become tougher for growing. As one of the report's authors noted, "Food security will be increasingly affected by future climate change through yield declines—especially in the tropics—increased prices, reduced nutrient quality, and supply chain disruptions."[4]

Climate change already impacts how we grow our crops. Some soils are becoming differently suitable for specific crops—some more, some less. Earlier warm seasons and later cold seasons have changed how fickle crops can respond throughout their respective growth cycles. Rapid changes between hot and cold spells with increasingly erratic periods of rainfall are already wreaking havoc on temperature-sensitive crops like coffee and cocoa. Farmers around the world accustomed to cultivating rice have switched to ancient grains more known for thriving in the arid, parched soils of the ancient world. Millet, for example, is now increasingly booming in India because it will grow in extreme heat and dry conditions.

What can we possibly do?

To start: what if I told you that many "farms" of the future will be able to decrease water use by up to 90 percent, eliminate the need for pesticides, grow larger and more nutrient-packed crops, be freed from soil constrictions like moisture and excess nitrogen and nutrient depletion—while also running on renewable energy?

Such are the promises of advanced next-gen agricultural technology. Or, "ag-tech" for short.

Before we discuss farms of the future at a civilizational scale, I should note that more and more new technologies are being developed to aid traditional farmers, as well. Brand new innovations in "agricultural intelligence" exist for sensing water levels in fields and fine-tuning how water is deployed throughout crop zones. Farmers can use sensors for determining discriminate locations for pesticides. So-called "ag-bots" can now be deployed to gather data on crops, including drone-scans for 3D mapping, seed-dispersal and fertilizer data to be gathered for making more informed farming decisions.

Specialized drones can even micromanage individual crop rows using cameras. Companies like Sunnyvale-based Blue River Technology (acquired by John Deere) uses a "see and spray" technique of identifying weeds for autonomous herbicide-targeting, reducing the use thereof up to 90 percent.[5] Other companies, like France's Naïo Technologies, utilize a series of vegetable-, fruit-, or vine-based, AI-guided weed management bots able to discern crop from invader and dispatch accordingly.[6] All of these techs and more are revolutionizing required amounts of energy (provided these bots are renewably-powered), nitrogen, and water, as well as whether pesticides are required.

As much as we may improve our technical capabilities, scientists and engineers are spending just as much time working on improving our crops themselves. To say nothing of how advanced new science is being applied to our processed foods, our fast foods, and our global meat industry. Our newfound capabilities in bioengineering will also factor heavily into decarbonizing our foods while still managing to feed the world.

Public opinion polls tend to show our populations are still fairly divided on the consumption of genetically modified organisms (GMOs). Which is unfortunate, given the vast new potential for improving our foods simply waiting for a legislative go-ahead. Regardless, scientists have still managed to find ways of improving what we eat entirely without direct genetic engineering.

When you think about it, we humans have been directly inserting ourselves into the evolution of our crops for thousands of years already. With every seasonal harvest, farmers intentionally disperse the seeds of sweeter apples and redder tomatoes while casting aside the others. Generation after generation of this process synthetically directs evolutionary adaptation toward the real selection at work: our own.

Fortunately, we are now much more skilled in analyzing different aspects of how crops grow—with unique new methods for identifying ways in which generations may have beneficial traits. Scientists can essentially keep track of crop genetics between generations and then run individual strains through an assembly line of sensors to measure every stage of development. Brand new speed-breeding and high-tech crop monitoring installations have changed the game in supercharging the normal, non-GMO development of vital crops. Rather than waiting for long periods of crop generation growth, scientists are able to grow thousands of crops at once and utilize the help of AI and machine learning to assess the traits they need—before hybridizing where needed.

Directed evolution in cross-breeding potatoes from around the world, for example, has already yielded super potatoes fortified with more iron and zinc to help reduce malnutrition in developing countries. The super-spuds should be available in 2021.[7] So our hands are certainly not tied with anti-GMO laws. Scientists can basically operate from a "genetic wish list" search criteria, scouring the world for any crop variants that may display advantageous genes along a number of measures. Sometimes they're looking for crops that grow larger, produce higher yields, or maybe faster yields; or perhaps crops that are more flood-tolerant or even salt-tolerant.

A company like Cambridge-based Inari can accomplish all of the above through its "Seed Foundry"—studying the development of multitudes of crops from seed to fully-grown plant. Inari's facility packs a dizzying number of data points to track all aspects of a plant's lifecycle across a number of different variables to assess what will impact its development. Inari can then isolate genetically advantageous features along different variables for further development.

By combining the use of machine learning and deep analytics with the different variables impacting plant development, Inari has made some interesting observations. Take soybeans, for example. Different areas of potential genetic influence include the number of bean pods per node or the number of seeds per pod. More pods can mean more yield for a marketable product. However, too long of an internode length can mean that plants will grow in ways that make them more difficult to harvest (leading to lost crops thereafter). Precisely generating the genetic assessments of what produces these states may have taken us generations before. Indeed, it may have taken us generations to produce redder, sweeter tomatoes—but Inari has been able to tailor tomato plant seeds to alter the size of their branches, the number of their flowers, and fruit size expansion thereafter. All of which in tandem have increased productivity more than 90 percent, according to company CEO Ponsi Trivisvavet.[8]

Work like Inari's could be invaluable. After all, you only need to improve a seed once. Then you have supercrops essentially forever. And surprisingly, an important part of Inari's work is also ensuring that crops display more genetic biodiversity. We have decimated crop biodiversity through monocultures and selective breeding before—forces putting our global food supply at significant danger if a crop disease ever spreads internationally. So any campaign aiming to increase the biodiversity of crops we depend on is great news.

Academic papers published by Inari's co-founder Steve Jacobsen outline how genetic editing tools can be used for "epigenetic" modifications like genetic silencing rather than editing. By using a tool like CRISPR and similar gene editing techniques, researchers are able to remove the DNA methylation that could otherwise suppress genetic activity—leading to a gene being more activated. Epigenetics is much like normal genetics, but focuses more on how genes are *expressed* than what they actually *are*. Consider sheet music: Studying the notes on a piece of sheet music is the field of genetics we know and love. Epigenetics focuses more on how the notes are played.

For centuries, we have been resigned to direct evolution by waiting for an advantageous crop mutation to present itself—and then just being smart enough to protect and foster it. This has been a long, painstaking process, often taking generations to produce substantive changes (and even then, with no guarantee of success). Normal evolutionary mutations are random and may take many years to present themselves.

Scientists in the twentieth century were able to speed this process up by introducing mutations synthetically. They bombarded seeds with chemicals and x-rays to induce mutagenesis through radiation. By using these methods, scientists were able to introduce several advantageous traits into various fruits and vegetables faster than they may have shown up otherwise. They were able to reduce a timeline of "potentially generations" to a decade or less. New techniques like CRISPR offer us the ability to turn decades into weeks.

Unlike bioengineering methods such as dampening certain genes or removing others, outcry against GMOs has typically focused largely on what are called *transgenes*. These are genes spliced from one species and simply handed to another. In other words, a genetic outcome far less likely under traditional Darwinian evolution. These are the so-called "Frankenfoods" you've probably heard about (and may be opposed to). But for the most part, scientists just want to edit crops in simplistic ways that will make them more available for people who need them.

With the power of CRISPR, we can make crops bigger, grow faster, produce more nutrients—and be more resilient to pests, droughts, and higher temperatures. With more people to feed in a warming world, the "GMO debate" about our methods of producing superior crops may ultimately end for practical reasons. No surprise that China, with the world's highest human population, supposedly has more than 20 labs dedicated to developing CRISPR crops.[9]

How, exactly, might genetic scientists go about increasing crop yield? Sometimes, this involves engineering how developing crops utilize photosynthesis itself.

One research team was able to hijack crucial parts of the photosynthesis process to increase crop growth an astounding 40 percent.[10] The team's insight was that most of our plant crops actually suffer from something of a photosynthetic glitch. As a result, plants employ a rather cumbersome chemical process of photorespiration to bypass this glitch, which takes a complicated route through several compartments of the plant cell. This process ultimately lowers a crop's prospective yield overall. But crops engineered to bypass this route with a photorespiratory "shortcut" can reportedly be 40 percent more productive, even in real-world conditions replicated over a two-year span. According to researchers, "We could feed up to 200 million additional people with the calories lost to photo-respiration in the Midwestern U.S. each year."[11] Crops able to potentially receive this treatment include rice, potatoes, soybeans, tomatoes, eggplants, and more.

These researchers are not alone: an international project involving scientists from around the world called Realizing Increased Photosynthetic Efficiency (RIPE) is working to increase other aspects of photosynthetic efficiency. With support from the Bill & Melinda Gates Foundation, the project aims to help improve crop yields to feed more of humanity. Which again, may be integral as temperatures get hotter.

Plant enzymes already have difficulty fixing CO_2 from the air; hotter temperatures tend to increase this difficulty. Hence one project that RIPE scientists are working on is hunting for faster and more accurate versions of this process already active in other plants and crops—to reintroduce in more that would benefit from the change. Researchers estimate that while the technology is five to ten years away from widespread use, that's really when we'll need them. As a philanthropic effort (thank you Bill & Melinda Gates), farmers of Sub-Saharan Africa and Southeast Asia specifically will enjoy royalty-free access to the spoils of this research.

Convergent research happening in other parts of the world has also focused on photorespiration. One team found that presenting rice with a photorespiratory bypass could increase yields by up to 27 percent. Engineered plants could again grow greener and larger with increased photosynthetic efficiency under field conditions.[12]

Other genomic research aiming to improve the process has focused on the transferability of more advantageous forms of photosynthesis. For example, did you know that different crops and plants use slightly different variants of the process? Crops like sugarcane, corn, and sorghum use so-called "C4 photosynthesis" (in which "C_4" refers to the four-carbon molecule first produced from this style of carbon fixation). Many other crops use C3 (carbon molecule C_3) instead, due simply to the environmental pressures favoring particular pathways in different climates. We can potentially "upgrade" C3 plants to C4 and improve photosynthetic energy conversion rates in warmer climates by inserting genes of plants more efficient in the process.[13] The change would not be more effective in all cases, of course, as C3 plants tend to be more productive in cooler climates. But in some cases, the impact could be very advantageous.

In another study led by scientists at the University of Sheffield, researchers managed to isolate a protein complex design that significantly influences plant growth throughout photosynthesis. Altering the structure or the expression of that protein complex could fuel investigations into whether there are more advantageous forms of it for growing bigger plants. Of course, photosynthesis is a marvel of nature's engineering prowess. But as with any blueprint created by the halfway scaffolds of evolutionary chance, it is far from perfect. In the words of the Sheffield research team's work regarding larger crops and more food:

Previous studies have shown that by manipulating the levels of this complex we can grow bigger and better plants. With the new insights we have obtained from our structure we can hope to rationally redesign photosynthesis in crop plants to achieve the higher yields we urgently need to sustain a projected global population of 9-10 billion by 2050.[14]

Or how about modifying crops to require less water? Researchers submitting work to the UK government have investigated the possibility of growing crops in saline (read: salt) water through a variety of genetic editing and selective breeding. If salt-tolerant breeds require less fresh water, that means freeing up more drinkable water for other uses. As a result, scientists have found they can increase grain yields on saline soil (where only smaller percentages of yields could be obtained with such difficult water supplies before).[15]

By using advanced genetic analysis and editing technology, research teams have powered ahead with other crucial crop developments. In just the past eight-or-so years since the advent of CRISPR and related tech, scientists around the world have already been able to create higher-yield strains of corn, soy, and wheat. But in addition to that, they have brought the world:

- Fungus-resistant bananas.[16]
- So-called "jointless" tomatoes able to grow larger and detach more easily for harvesting.[17]
- Strains of tomatoes outfitted with sweet pepper genes which defend against bacterial disease[18] and other tomato strains with roughly 5 times the antioxidants of cherry tomatoes.[19]
- Non-browning mushrooms.[20]
- Advantageous forms of starch in cassava.[21]
- Fire blight-resistant Gala Galaxy apples.[22]
- Hypoallergenic wheat.[23]
- Rice engineered to use just 60 percent of the water normally required for growing, under higher atmospheric levels of CO_2, able to survive drought and in environments with higher temperatures.[24]
- Virus-resistant papaya.[25]

There's also herbicide-resistant wheat, which allows farmers to kill weeds indiscriminately while sparing valuable crops. That's bound to be controversial, but in the meantime it spares crops from competition with weeds for sunlight and soil resources while speeding up the harvesting/release of crops.[26]

Or how about carrots with increased levels of bioavailable calcium? (All right, these were introduced previously).[27] As well as folate-fortified lettuce,[28] rice with increased iron, zinc, and beta-carotene (precursor for vitamin A).[29]

Synthetic biologists have even used advanced genetics to prospect an increase in crops' immune resilience to environmental stressors of all kinds.[30] Sounds like a desirable trait if we want more healthy harvests, doesn't it?

Genetic engineering has also allowed researchers to introduce insect-repellant (or even outright insecticidal) traits into crops that are dangerous for pests but not for us. These chemicals operate similarly to how chocolate is dangerous for dogs, but not for humans. And this newfound capability in targeting chemicals has opened up entirely new pesticide-free methods of deterring pests.

Insecticidal maize, cotton, and poplar (among others) have all been produced already.[31] Or consider perhaps the most famous among insecticidal crops: Bt eggplant strains in Bangladesh. These are strains of eggplant engineered to enable the *Bacillus thuringiensis* (Bt) bacteria to produce a protein toxic to pests, but not for humans, within the crops themselves. It was officially the first genetically edited food to be commercialized in South Asia.[32]

Pest control through biological augmentation will undoubtedly be a huge hit with farmers. After all, you can save money and can market pesticide-free food to customers. Moreover, losing fewer crops to pests certainly saves on the energy and emissions of growing more. Which may not sound like a huge deal right away—but by some estimates, the monetary penalty of crop loss alone comes into the hundreds of billions in lost dollars worldwide.[33]

Canadian company Terramera, on the other hand, has augmented pesticides with naturally insecticidal chemicals found in certain kinds of trees. Terramera took the Indian concept of using tropical neem tree extracts as natural bedbug repellants and isolated natural ingredients for deterring crop pests in general—which they then diluted into normal pesticides. The result is fewer chemical concoctions in pesticides while offering similar amounts of crops saved from pests.

Scientists are continually studying the ways that wild plants deter their own pests, often with the aim of plagiarizing those chemicals for ourselves. And ultimately, a budding new "biopesticide" industry is expected to surge in the coming years as bioengineers become better at fine-tuning naturally produced, pest-repellant gene activity.

This is the brave, new agricultural world in which we find ourselves. A company like Caribou Biosciences, mentioned earlier—whose slogan is *Engineering any genome, at any site, in any way*,[34] furthers crop research in producing advantageous genetic traits for select foods. From their own site: "Caribou is pursuing accelerated precision breeding strategies to speed up and streamline the processes for the generation of new traits such as drought tolerance and disease resistance to protect plant health, increase crop yields, and develop healthier crops." And with more than 20 new patents in 2019 alone, the race is on for biotech companies to discover the most beneficial new food traits sooner than later.

On top of re-thinking our pesticides and how crops grow, incredible new academic work is being done to re-engineer another crucial aspect of our food system: nitrogen.

Nitrogen is an important fertilizer that helps us sustain larger crop yields to feed larger populations of people. Plants have normally obtained their nitrogen through the microbes in our soils that produce nitrogenase—an enzyme that fixes nitrogen from the atmosphere and converts it to ammonia. Plants then take in this chemical through their roots. Blanketing crops with synthetic fertilizers certainly gives us more crops, but it has dampened microbes' natural ability to do the above.

By some estimates, if we were to cut out fertilizers, the world's food production would be cut nearly in half. So we clearly need fertilizers, and therefore chemicals like nitrogen. But we also need to avoid the problems associated with how we currently create and use various fertilizers.

Producing a chemical like ammonia consumes an incredible amount of the world's energy. We require energy-intensive mining to acquire phosphate. And then, when only about half of nitrogenous fertilizers applied to fields are taken up by crops,[35] a portion is lost as gaseous nitrous oxide (N_2O)—a greenhouse gas 300 times more powerful than CO_2 and responsible for 5 percent of worldwide GHG emissions.[36] Worse yet, new research from scientists across Europe and the US tells us that N_2O

emissions have risen steadily over the past two decades, coinciding with increases in synthetic fertilizer use.[37] So fixing our nitrogen problem before we go ramping up food production would spare us from critical greenhouse gas emissions.

But aside from excess nitrogen, agricultural fertilizer also releases immense amounts of methane. A June 2019 study from Cornell University revealed that fertilizer plants emit *100 times* more methane than previously reported.[38] You will recall that methane itself is much more potent a greenhouse gas than CO_2. So there are clearly significant problems with the production and dissemination of synthetic fertilizers in agriculture. Managing how we produce and use our fertilizers will therefore be paramount in cutting emissions from agriculture.

One advanced new solution for our nitrogen problem is to replace synthetic, energy-intensive nitrogen production with on-site, renewably produced fertilizer substitutes. Which is to say: empowering processes that lead to the soil producing its *own* nitrogen at levels plants will need and not much more. This has been made possible by the use of an engineered cyanobacteria (*Xanthobacter autotrophicus*) in the lab of Harvard's very own Daniel Nocera—a name you will remember from the discussion about artificial photosynthesis.

As with photosynthesis, the technology of Nocera's new lab splits H_2O into H_2 and O, but then takes that hydrogen and feeds it to the engineered bacteria. Liquid solutions containing the bacteria are then sprayed onto fields. When the solutions reach fields, airborne nitrogen can be fixed by the microbes to create ammonium—which plants then store in their roots and use as needed. In this way, plants can essentially be handed natural nitrogen in the amount(s) they require rather than blanketed with an excess in synthetic forms.[39] On top of that, the approach eliminates the run-off that produces marine dead zones. And, as an added bonus, the bacteria absorb small amounts of CO_2 from the air.

Together with colleagues Pamela Silver and Xiaowen Feng, Nocera has formed Kula Bio with partners Bill Brady and Kelsey Sakimoto. The Kula team hopes to commercialize their run-off resistant biofertilizer as a low-cost and more sustainable alternative to traditional fertilizers. And early results tell us the approach has promise: by using their more sustainable biofertilizer, the lab was able to grow radishes more than three times the size of control group radishes.[40] All without the energy or carbon expenditures going into synthetic fertilizer refinement.

A company like Berkeley-based Pivot Bio (another start-up partially funded by Gates's Breakthrough Ventures) has another incredible nitrogen solution. Engineered microbes are sprayed onto soils as above, but they then attach to a crop's roots where they colonize and fix nitrogen for direct uptake. Pivot views their approach as replenishing microbes' natural ability to create a nitrogen feeding loop for plants lost after the advent of synthetic fertilizers.[41]

Boston-based Joyn Bio also enjoys elite-level funding from parent companies Bayer and Ginkgo Bioworks. Joyn uses genetic libraries of dozens of microbes capable of colonizing corn or wheat plants with the same kinds of applications. The result, again, saves the energy and emissions of producing the same crop outcomes the now-traditional way. With applications like these, farmers will not need expensive, synthetic fertilizers, nor harsh chemical pesticides.

Boston-based Indigo Ag is another impending ag-tech giant creating designer microbes capable of fixing nitrogen naturally. Landing at the top spot overall on CNBC's 2019 "Disruptor 50" list—an inventory of companies revolutionizing their respective industries—Indigo Ag also specializes in producing heat and drought-tolerant strains of crops like corn, cotton, soybeans, and more. Indigo is currently, at this time of writing, a $3.5 billion company in the US, indicating that next-generation ag-tech is very well positioned to take over the market.[42]

Even more recently, scientists at MIT have designed specialty "seed coatings," transferring large benefits to fledgling crop seeds. MIT's coatings allow new seeds to more readily acquire natural nitrogen from soils (again saving on synthetic use) while also providing seeds the necessary nutrients they need to grow in otherwise less hospitable soils.[43] The seeds are coated with a silk treated with nitrogen-fixing bacteria—that also allows for increased growth in salty soils. As a result, these coatings potentially open up many more places to plant crops. The team's researchers believe that future coatings will focus on gathering moisture from drought-stricken soils to open up even more crop land. Although research into the impact new "agricultural frontiers" will have on biodiversity also exists, so scientists are still debating on where and when they may enable new farm grounds.[44]

When we consider the impacts a changing climate may have on our crops, tougher and stronger traits may well be necessary. For example, flooding already destroys 4 million tonnes of rice every year in countries like India and Bangladesh.[45] That lost rice constitutes meals for about 30 million people. New DNA-assisted breeding techniques can and have produced "submergence-tolerant" yields to surpass this danger. So while the developed world may sometimes balk at engineered crops as undesirable, generating crops on our genetic wish list may simply be crucial for a world that soon needs them.

Elsewhere in the world of ag-tech, rather than dreaming up new strategies to augment the world's existing agricultural infrastructure, some scientists are focusing on radically re-defining the entire edifice of how we grow and transport food. Consider, if you will, the latest innovation in agricultural science: indoor vertical farms.

INDOOR VERTICAL FARMS

The idea of growing our food indoors—in what can only be described as "plant factories"—may not seem like an immediately intuitive concept. After all, sunlight is free. Plants have been growing outdoors since the beginning of plant life on Earth. And if indoor operations need to be climate-controlled and illuminated with artificial light, just how exactly can this operation scale to feed large populations of people? In fewer words: how is this not insane?

For starters, many indoor vertical farming companies have repurposed urban spaces very close to where food will eventually be going. This eliminates transportation costs and the crop waste involved. Trucks are not transporting crops hundreds or even thousands of miles from farm to plate. Controlled environments also eliminate the need for pesticides because there are no pests inside. So cut out the cost of pesticides and increase the draw for selling produce without them. Contaminants like E. coli do not present a danger. Vertical stacks of crops can save on space and the costs of heavy equipment needed to traverse large amounts of land—no tillers or tractors needed. Water is conserved by up to a stunning 90 to 95 percent in most facilities, with nutrient-filled water misted or washed over roots and then recycled for future use. Thousands of strategically placed micro-sensors linked to cloud-connected AI can assess every aspect of every crop's growth-cycle, reducing the need for large numbers of staff to monitor.

Indeed, AI sensor systems produce real-time data for indoor operations tracking everything from moisture levels to temperatures, light-exposure levels, and more. And it might sound surprising, but altering growth variables like day/night light cycles, moisture applications, nutrient contents, and more can produce highly variable (but predictable) outcomes in plants. Altering a plant's light/dark cycle can make it grow in specific ways. As can altering when and how plants receive water or nutrients. New variations in growth cycle variables have produced interesting outcomes for indoor crops, including taste, color, mouthfeel, and more. Signature light and nutrient recipes allow for incredible predictability in overall crop yields. As a result, vertical farm operators can essentially tailor the amounts of sunlight, rain, and wind reaching crops, removing a large amount of crop yield uncertainty. Moreover, specially placed bots can assist at all stages of these processes to create a high level of harvest automation.

Now, imagine an indoor facility's lights, sensors, and crop-tending bots all powered by renewable energy to grow food at literally any time of the year, anywhere in the world, with faster growth cycles for more potential harvests. Still sound crazy?

There are now several varieties of indoor farming operations popping up around the world. Hydroponic, aquaponic, and aeroponic operations all offer various benefits for indoor farmers. Essentially, none of these options use soil—instead relying on nutrient-rich water to provide sustenance for plant roots. Aquaponics involves more of an ecosystem approach with fish providing a role in the nutrients eventually used by roots. Hydroponics simply mixes nutrients with the water in which plant roots are soaked. Aeroponics uses neither, instead misting plant roots suspended in the air (which can be surprisingly good for some roots with additional oxygen exposure resulting in faster growth).

Consider an aeroponics operation like New Jersey-based AeroFarms. Aero is an exciting company for two reasons: commercial-scale operations spanning four continents with installations operating all year round, and a signature use of aeroponics reducing water use by an incredible 95 percent compared to traditional agriculture. Aero's water system keeps nutrient mists in a closed-loop system where they can be reused over and over, while also following unique timing schedules. Rather than just once or twice a day, for example, crop roots are misted continuously throughout each day.

Nutrients within the applied water can also be tailored at different points of crop development. These differential applications will result in signature growth profiles for each crop that can be set in advance. Profiles may include altered nutrient contents and levels of magnesium, iron, zinc, and other mineral elements—all according to AI-assisted outcome productions for various crops. Even the cloth medium AeroFarms uses for seeding, germinating, growing, and harvesting is used from post-consumer recycled plastic (taking another estimated 350 water bottles out of the waste cycle for each indoor facility with cloths being reused).

Aero also boasts a stunning number of "data points" to assess plant growth throughout each harvest cycle. By cataloging different aspects of crop growth—like light levels, nutrient changes, misting schedules, and more—machine-learning algorithms can create surprisingly complex crop profiles. Even signature "light recipes" can be used by tailoring wavelengths and exposure times. After all: perhaps a crop grows better without the traditional 10 hours of darkness while waiting for the sun to come up—but with 10-minute sun/darkness intervals instead?

Aero's ability to micro-manage the growth cycle of a plant with granular precision has allowed for significantly faster harvest times. By their own estimates, they require essentially half the time as traditional farming.[46] Moreover, because the full seed-to-harvest cycle involves 130,000-plus data points, manipulating precise algorithms of light or nutrients actually allows Aero to modify the size, color, texture, shape, nutritional content, and flavor profile of foods grown.

Better yet: Aero's farms are being strategically located along major distribution routes and near population centers to take advantage of supply chain weaknesses. The team has repurposed locations like former nightclubs, a former laser tag stadium, and a former paintball arena. Aero's global HQ in Newark is the world's largest indoor vertical farm, at the time of this writing. Strategic placement slashes the emissions cost of transporting food from its source to point of sale. By Aero's analysis, roughly 98 percent of transport emissions typically produced by transporting food around our countries can be eliminated.[47]

Of course, there are many other aeroponic and hydroponic vertical farms out there. Richmond-based Airponix uses a "nutrient-laden fog" to create a cloud of <20-micron water droplets. Doing so allows them to conserve energy. An electrostatic charge causes those droplets to adhere to the roots and leaves of suspended plants directly, supposedly increasing nutrient uptake.

Even more impressive, indoor grow-op Crop One boasts using only

1/2500th of the water in a soil-based operation.[48] The energy and emissions in piping and channeling all that freshwater to traditional farms can apparently be reduced significantly.

French aeroponic start-up Agricool, like the aforementioned Aero-Farms, offers hyper-local potential with shipping-container-sized indoor farming modules. Their reconverted shipping containers offer all of the previous accoutrements—soil-free, nutrient-laden mist with light recipes in effect, etcetera. But the added bonus is that they can be adapted to any environment on short order for restaurants, hotels, and more—all while rigged to small-scale renewable energy inputs to power the containers' devices.

By packing their tech into tinier spaces, Agricool reports producing an average of 120 times more produce per square inch than traditional soil-based agriculture,[49] which is great news for the prospects of its larger installations. One facility slated for Dubai will reportedly rise to an estimated 130,000 square feet. Agricool containers can produce various crops, though their strawberries getting the most press thus far. Having approximately 30 percent more vitamins and 20 percent more sugar than conventional berries will garner that kind of attention.

Similarly, Beijing-based Alesca Life is pursuing the hyper-local, small-scale approach. Alesca's module-based "container farms" are designed to be embedded into buildings like schools, hotels, or other urban spaces. Urban mini-farms producing local celery, kale, herbs, and more have been found in the abandoned car parks of downtown Beijing.

Alesca uses a cloud-based connectivity system linking real-time IoT plant monitoring with user management controllable via app. Which means that users are able to follow light-and-nutrient recipes for optimal growth yields or reportedly craft their own. So if you're feeling creative, you might consider the interesting traits you could tailor for personal taste. And yes, the installation can even ping your smartphone when your spinach is ready for harvest.

Despite the benefits, however, cost/revenue margins for indoor farms can be slim until operations are scaled significantly. Growers therefore have a powerful incentive to automate as many aspects of farming and distribution as possible. This is why the robot-powered indoor vertical grow op in Pittsburgh—called Fifth Season—provides a compelling case. The company has managed to automate seed-planting within trays and the delivery of trays to grow-rooms—where autonomous systems control essentially every aspect of crop development. That includes light, CO_2 exposure, nutrient mix, and more. A series of sensors automates system awareness of crop readiness, at which point even harvesting is automated. And because live humans are essentially left out of this process, aisles for crops are much smaller than in other ops (allowing for more crops over-all). Automated systems even take care of many packaging steps and grow-tray sanitizing to start the process anew. San Francisco-based Plenty also boasts many of the same accoutrements: microsensors everywhere, full automation, massive aisles of robot-assisted growth, and harvest cycles for stacked rows of crops.

Morehead, Kentucky-based AppHarvest has integrated an impressive number of renewable inputs. The operation uses a greenhouse roof to help incorporate natural sunlight when possible—while also retaining rainwater via rooftop retention ponds to help off-set even more water use.

Indoor and vertical grow-ops also may increasingly utilize bioengi-neered foods more suitable for those growth mediums. Consider that sci-entists have developed a CRISPR'd tomato plant with shortened vines and clusters of fruit more resembling grapes.[50] The alteration will allow tomatoes to grow in uniquely challenging locations. And while researchers themselves were picturing space-travel enclosures, we're thinking about vertical stacks here. After all, AI-assisted harvesting tech can far more easily pluck grape-like tomatoes right off the vine over traditional arrangements.

Research teams have also found ways to speed up the growth process in various crops to produce more fully grown plants per harvest cycle. I mentioned AeroFarms' precision approach to micromanaging growth cy-cles to create faster harvest times. Recent research in academia has also bolstered the concept of adjusting light-cycles for vastly increased growth speed. In yet another technique originally inspired by NASA research, crops can be tricked into flowering early if they receive sustained expo-sures to blue and red LED lights in cycles for 22 hours per day, tempera-tures relatively steady at 62 to 72°F. Scientists following this recipe revealed that they could grow up to six generations of wheat, barley, chick-peas, and canola in a year, whereas traditional growing methods would produce only one or two annually.[51]

Of course, indoor vertical grow-ops are not without their drawbacks. At this point, they do not scale to feed civilizational levels of consumers. They also require grid power should renewable inputs drop out—which is important to link, as exposed root systems in aeroponic operations can desiccate without continual misting. Sensors also cannot mismanage crop nutrient levels, given the financial risk for growers potentially losing a harvest. These concerns are not trivial; even small issues like mineral deposits on misters due to vaporous water contact can be an ongoing menace. Indoor vertical grow-ops are therefore clearly a solution in progress, and just one of many to consider.

MEATING OUR CLIMATE GOALS

Clearly, we need to re-think how we grow our food. But beyond fruits and vegetables, other aspects of our food production system carry oversized energy and carbon emissions costs. Producing meat—essentially *any* kind of meat—carries a particularly huge footprint in terms of energy cost and carbon emissions. Fortunately, advanced new sciences and technologies can again lower both to sustainable levels.

By now, you've probably heard that meat production is an incredibly wasteful use of our land and poses numerous environmental problems. A growing number of meat-eaters worldwide will make these problems worse. We hear these points on the news and on our social media feeds; from our friends and our favorite celebrities and artists. In fact, as I was writing this section, pop stars Beyoncé and Jay-Z were offering their fans free tickets for life (or 30 years, anyway) in a campaign to get more people onto a plant-based diet. Events like "Veganuary" were just picking up steam and more and more people were trying out "Beyond Meat" burgers at local fast food chains.

Many of us consume far more meat than we need to. Which, in terms of the resources going into producing our calories, is spectacularly wasteful. Every piece of meat we consume necessitates that land has been used somewhere to grow and care for the animal it came from. It involves the footprint of their housing and well-being, the processing and shipping of those animals and keeping everything cooled the entire way.

Every aspect of meat production carries its own associated energy and emissions cost. As Bill Gates noted on his blog back in 2018: if "cattle" were its own country, it would currently rank third overall in carbon emissions.[52] Resource-wise, depending on the animal, producing one pound of meat protein with Western industrialized methods requires 4 to 25 times more water, 6 to 17 times more land, and 6 to 20 times more fossil fuels than producing a pound of plant protein.[53]

We also sacrifice the land being used to raise animals for meat consumption. This is land often cleared for livestock owners through a razing of forested areas crucial for our planet's health. In fact, Project Drawdown Executive Director Jonathan Foley cited tropical deforestation—often undertaken so land prospectors can raise livestock or grow crops to feed other livestock—as potentially the number-one source of greenhouse gases from land use and agriculture.[54] Which makes sense, if all the carbon stored in trees is simply being released back into the atmosphere.

Only about 18 percent of our calories come from meat and dairy—even as those two calorie sources take up 83 percent of our farmland.[55] Incredibly, almost 30 percent of the world's ice-free land is now used to raise livestock.[56]

It should be no surprise, then, that given our energy costs associated with meat—and given our carbon costs especially—academics worldwide are calling for us to reduce meat consumption as much as we can. One letter signed by more than 50 leading scientists published in *Lancet* recently argued that the world must reach "peak meat" by 2030 (in that meat production cannot increase any further at that point) or we will face significant consequences.[57] The IPCC's most recent sustainability report declared that shifting from meat-heavy diets will be crucial if we want to limit global warming to 1.5°C.[58]

If you're like me, however, you love meat—or, at the very least, the taste of it. And while I'm prepared to give up more of it for the climate, there are fortunately—finally—several compelling alternatives poised to reduce the footprint of meat.

To start, we have food scientists working on incredible new means of simulating the tastes of meat to mix with plant-based alternatives. These are the plant-based meat products you now see in fast food chains and supermarket deli aisles. Only through painstaking scientific efforts have these products been made possible—and they stand to improve as bioengineers better understand the molecular components involved. We also have researchers investigating entirely new sources of protein for those folks who want to ditch the meat but keep the protein. In what should have been a headline to break the internet in 2018, scientists have now managed to create protein from water, sunlight, and *air* thanks to research originally spearheaded by NASA. We also have synthetic biologists working on "cultured cells" to grow meat tissue without a full animal—and without the full energy or carbon footprint. All of these developments stand to radically revolutionize how we produce and distribute meat.

As we wait for these options to improve and converge on our lives, there are actually several steps scientists are taking to reduce livestock emissions in the meantime. For example, you have probably heard that "farting cows" or burping livestock are a significant source of greenhouse gases. And it's true; they are. Ruminant livestock (cows, sheep, goats) have evolved to process plant food with a special digestion process that produces large amounts of methane. These releases of methane are estimated to account for about 6 to 7 percent of annual greenhouse gas emissions worldwide.[59] That number will only grow as more of the world has the choice of meat for dinner. By one estimate, population levels in the year 2050 will consume roughly 70 percent more meat than we did in 2005.[60]

New research tells us that simply feeding our livestock seaweed can vastly reduce this livestock-emissions number. One strain of red algae, *Asparagopsis taxiformis*, was found to reduce methane production in ruminant livestock by up to 99 percent—and it only needed to constitute about 2 percent of our livestock's feed supply.[61] Scientists are also investigating whether a small group of gut microbes may be responsible for all the methane—and whether they are able to be silenced through CRISPR editing.[62]

Fortunately, plant-based meat alternatives do already exist. Many are not significantly healthier than traditional meat, but the reduced carbon footprint is a big improvement. Impossible Foods is a name you may recognize—the California-based company famously crafted the Impossible Burger, a plant-based alternative with a surprisingly similar look, feel, and taste.

Impossible was able to blend a melange of soy proteins to mimic the flavor, aroma, and texture of beef. A "culinary binder" called methylcellulose improves the texture. The burger even bleeds like real meat, courtesy of an ingredient from soybeans called leghemoglobin, a protein bound to another molecule called heme. You might recognize heme from the hemoglobin in blood, which helps to regulate the oxygen supply in cells. As a result of incorporating heme molecules—which give meats their red color—the Impossible Burger can approximate the look and sensation of real meat.

The type of leghemoglobin used in Impossible's burgers does not come directly from soybeans, but from a strain of yeast engineered with soybeans' genetic instructions to produce the chemical instead (undoubtedly to save time and money). The US Food and Drug Administration has therefore deemed Impossible's use of leghemoglobin "generally recognized as safe" (GRAS), but the alteration does technically make the product a GMO.

By Impossible's own estimates, producing their burger requires 87 percent less water than traditional burgers, generates 89 percent fewer greenhouse gas emissions, and uses 96 percent less land.[63] Great news, then, that the market now enjoys several competitors in the plant-based-alternative arena. Beyond Meat, maker of the Beyond Burger, was formally assessed by the University of Michigan. The results indicated that Beyond uses roughly 99 percent less water, 93 percent less land, 46 percent less energy, and produces 90 percent fewer greenhouse gases than conventional beef burgers.[64]

It might surprise you to hear that some of the largest producers of meat-less meat are large-scale meat producers themselves. Mega-meat producers in America like Tyson, Smithfield, Hormel, and others all have strong lineups of meat alternatives. Nestlé even answered with its Awesome Burger, an attempt to keep Beyond and Impossible from running away with the market.

It *is* important to note that an Impossible or Beyond burger is still a burger—highly processed, still high in sodium, and not likely to impress many nutritionists. But they do at least offer some of us an off-ramp from meat.

Now, this is where our meat-less alternative discussion gets truly sci-fi. Yes, it is great that we have scientists working to make traditional agriculture more sustainable, and that we are using biotechnology to create meat-like simulacra that can satisfy our carnivorous cravings more sustainably. But it is also fantastic—and just plain cool—that we have biochemists and geneticists working on developing cell cultures in lab environments destined to grow into meat tissue, without much of the required water and carbon emissions.

Cultured meat is what industry regulars now call "clean" meat. Meat has been dubbed to be clean when it is cultured in vitro—which is to say, grown and developed in a lab setting. Scientists already had this ability, but not nearly at the skill level and commercial-grade quality of what is possible today.

Cultured meat is essentially meat tissue grown outside of an animal's body. Scientists can take stem cells destined to become muscle tissue and place them in a bioreactor to undergo exponential cell growth. The reactor contains nutrients and all the other naturally occurring growth factors, promoting cell growth for the cultured batch. These reactors can often resemble the same kinds of fermentation areas where beer and yogurt are produced. When scientists then deprive the stem cells of growth factors, they know it's time to grow. Ultimately, cells then merge to form thin muscle fibers that look first like little tubes before they are placed into a watery gel that groups them together into the muscle fibers we recognize as meat. As these groupings of muscle cells contract, they begin to look more and more like typical tissue.

According to Dutch company Mosa Meats, from just one sample cow we can produce about 80,000 quarter-pounder-sized patties of cultured beef.[65] Better yet, for some who prefer it this way: it's all non-GMO. Mosa, like Berkeley-based Memphis Meats, are hoping for competitively priced products in later 2020 and commercial products shortly thereafter. So it's still a bit early to tell when cultured meat will end up on restaurant menus, but by some estimates it won't take long for them to dominate the market. Global consulting firm AT Kearney recently estimated that by 2040, most of the meat we eat will not come from animals.[66]

Companies like Future Meat Technologies, Beyond Meat, and Mosa Meat, backed by Google co-founder Sergey Brin, are just a few of the more notable businesses venturing into clean meat. Meanwhile, Israeli start-up Aleph Farms even claims to have created the world's first lab-grown *steak*. The team has more recently cultivated meat in space with the help of a 3D bioprinter.[67]

Unfortunately, cultured meat is not yet available for commercial purchase. Curiously, progress in bringing cultured meats to market actually seems to be happening more quickly in the seafood department. Emeryville-based Finless Foods has developed cultured seafood through similar feats of engineering along with Beyond Meat and Mosa. Finless Foods produces bluefin tuna without the associated fish farming, marine contaminants, or any of the associated industrial infrastructure. Among their end products is a burger-like shaped piece of bluefin tuna consumers can cook in a pan or stir fry.

San Diego-based BlueNalu also specializes in "cellular aquaculture"—effectively saving us from the emissions profiles of fishing fleets to catch fish, grow and monitor stocks, and process the fish thereafter. Shipping materials between those different stages—all gone. No hormones or antibiotics or massive aquaculture facilities.

Now, even if some may not believe that animal-free products can replace the originals—multinational investing firms sure do. The biotech industry is already moving beyond biofuels, brewed clothing, and eco-skis. New biotech applications in our food industries are also targeting designer molecular mixtures to improve sustainable foods. Notably, bioengineers are seeing remarkable progress in making animal-free food additives like vegan cheeses and milks more palatable.

For all the progress we make in replacing meat, industry critics still cite the need for dairy products. Vegan cheese just doesn't measure up, they say (and often rightfully so). But to believe that up-and-coming industry giants can never do better is silly. Consider that Ginkgo Bioworks has already branched into the food additives business with spin-off lab Motif Bioworks (yet another BEV-invested company). Like Ginkgo, Motif also uses bioengineered organisms to produce signature chemicals—these are just edible.

Motif begins with an extensive analysis of which food qualities tend to provide the sensory experiences we enjoy. There are specific chemicals involved in generating the smells, tastes, and mouthfeel of particular foods. They then decode the genetic makeup of ingredients that factor into providing us those experiences. The company follows by designing organisms with the genetic programming to assemble those same chemicals via fermentation. And before that starts to sound too unusual, one might remember that beer and wine are made the same way.

With the use of microbes, manufacturers can bypass the concerns of working with animal products. Manufacturers need not worry about processing, say, bones or fur, or caring for the well-being of animals who need to eat, sleep, and be housed adequately. As a result, labs like Motif stand a great chance at significantly reducing the cost of purchasing food additives major companies might be looking for. This again, along with meat alternatives and biotech fabric alternatives, reduces the need for livestock space.

It is true that we have had a variety of milk substitutes for years. Plant-based milks like soy, almond, coconut, and more are essentially produced by first breaking plant materials down. Then, their proteins are reconstituted in a water solution to resemble the same proteins produced in a lactating cow. "Resemble" is the key word here—the proteins still differ in crucial ways. Sophisticated tasters can often tell the difference between plant-based milk and cow milk. The difference-maker is that companies like Motif can now more closely resemble the real proteins.

Similar labs like New Zealand-based New Culture and Emeryville's Perfect Day have also analyzed the precise structure of non-dairy proteins for use in dairy. As it turns out, two of the proteins responsible for dairy's great taste are casein and whey protein. Both of these can now be produced by engineered microbes for integration into food additives.

New Culture has conducted double-blind tests with its lab-made mozzarella featuring microbe-produced casein proteins. The results, they say, have been very positive.[68] And the consumer data is encouraging for our prospects of taking the cows out of dairy: analyses from the Plant Based Foods Association and The Good Food Institute revealed that sales of plant-based ice cream and frozen desserts grew 27 percent last year, while plant-based cheese grew 19 percent and plant-based yogurt grew 39 percent.[69]

San Francisco-based Clara Foods is another American start-up focusing on animal-free food additives and alternatives. They call their approach "cellular agriculture"—and what they produce is chicken-free egg white proteins. Consumers can supplement these animal-free egg whites into recipes calling for eggs (or replacing the egg white portion, at least). That relieves, even if only slightly, the consumer demand for factory-farmed chickens. All consumers may not personally benefit from a sustainable product like chicken-free egg whites directly, but companies that require egg whites may enjoy a cheaper source of the material and fewer emissions tied to their end products.

Animal-free food alternatives for burgers, milk, cheese, and eggs won't be the end of sustainable options. For example, would you believe bean-free coffee? Seattle-based Atomo—located just down the street from the Pike Place Market of Starbucks fame—specializes in producing "molecular coffee." Atomo's product is a close chemical equivalent for traditional coffee blends, albeit without the use of industrially farmed coffee beans. Atomo gathers upcycled plant materials from other companies to create what they call a "carrier matrix" full of the necessary proteins, carbs, and oils required. This matrix of materials is cleansed and blended to provide the same flavor and mouthfeel as bean-based coffee. Blind taste tests have revealed that (in at least one campus, anyway) some tasters even chose Atomo's blend over the Starbucks Pike Place blend.[70]

If, like me, you sometimes wonder whether you are more coffee than human being, you are right to feel skeptical about molecular coffee. But when climate change is ravaging soils around the world, endangering our coffee and potentially making it more expensive, it may be crucial to keep an open mind. Clearly, we will need to consider new alternatives and new options in a more livable world. We will have to rethink how we use terms like "meat," "milk," and "ice cream" in the future. Maybe even "coffee." And even if all experts cannot agree that cultured additives will always be more carbon-free than traditional animal processing infrastructure,[71] it's hard to imagine how swapping out the animals, fields, and facilities for smaller labs couldn't be better in the long run.

RECONSIDERING SUSTAINABLE PROTEIN

Our efforts to replace products obtained from animals are ramping up quickly—which is great news for continually reducing how much energy we need to sustain ourselves, along with the carbon emissions involved. Some scientists have managed to develop even lighter, cleaner, and more sustainable protein sources than our pre-existing meat alternatives, deriving protein from sunlight, concentrated CO_2, and ambient air.

The tagline for Finnish food tech company Solar Foods, for example, reads: "food out of thin air." And while it's not quite that simple, the reality is still fairly close. The process again begins with renewable electricity splitting water's H_2O into H_2 and O. Hydrogen is then mixed with CO_2 and potassium, sodium, and a few other nutrients. This mix is then fed to proprietary microbe cultures, which use the hydrogen for energy as they consume CO_2. The resulting liquid produced is then dried into a powder with essentially the look and taste of wheat flour. Flour, that is, comprised of full or single-cell proteins and ready to supplement drinks and meals. The end result of the Solar Foods process is an edible powder they call Solein; roughly 20 to 25 percent carbs, 5 to 10 percent fat, and 50 percent protein.[72]

Incredibly, the microbial protein process was originally dreamt up by an engineer at NASA decades ago, but never pursued. Research teams envisioned that something of a closed loop could form between astronauts exhaling CO_2 for on-board microbe colonies to eat. Those microbes would produce chemicals to be made into protein in the process. Then astronauts would consume the protein and exhale more CO_2 to keep the process going. Solar Foods is in talks with the European Space Agency to market their technique.

For the rest of us here on Earth, a protein source like Solein is relatively independent of land constrictions, water limitations, and energy costs. Even populations in the Arctic or in the middle of deserts could produce sustainable protein. In fact, by their own estimates, producing 1 kg of Solein requires only 10 liters of water (as opposed to the 15,500 liters going into producing 1 kg of beef).[73]

The prospect of such a sustainable protein source generates a few new possibilities. A company like Solar Foods could market Solein to Impossible Foods, for example. The protein is suitable for essentially any diet, so could augment meal sources across the global food industry. Expect to see Solein commercialized sometime in 2021.

Another sustainable protein effort comes from Berkeley-based Air Protein, who also uses renewably powered electrolysis. The setup is essentially identical to Solar Foods, except that Air Protein (spun-out from growing sustainability giant Kiverdi) has also noted they may utilize captured carbon in the future to augment the operation. The land and resource savings are also equally incredible. Air Protein's CEO, Lisa Dyson, estimates that one could generate enough protein in a Disney World-sized facility to match a traditional farm the size of Texas.[74]

Green tech giant Kiverdi's research teams have also developed a new application for sustainable protein. Their so-called CO_2 Aquafeed is basically another version of Air Protein, but designed for consumption by marine permaculture fish. Normally, fish enclosed in permaculture arrangements feed on "fishmeal," a protein slurry derived from smaller forage fish. CO_2 Aquafeed, on the other hand, is derived from CO_2 and renewably powered electrolysis. This means those forage fish get to stay in their own ecosystems, permaculture operators get to save plenty of money, and the world sees a smaller carbon footprint. Sunnyvale-based Novonutrients operates much the same way in making a sustainable fishmeal. Their product, Novomeal, is produced by bacteria housed in bioreactors fed with industrial waste CO_2.

Developing these options may be more time-sensitive than we realize. One report from the UN released in August 2019 warned of "significant" disruptions to global food supply systems due to climate change. The report, which involved more than 100 experts from over 50 countries, revealed that a half-billion people are already living in places increasingly turning to desert. The soil in these regions is being lost 10 to 100 times faster than it is forming.[75] Researchers noted that even now, more than 10 percent of the world's population is malnourished. Declining crop yields could therefore deeply destabilize regions already beset with civil unrest.

It is also worth noting that research shows that climate change is boosting production of some food sources in greening locations due to warmer temperatures opening up new farming possibilities at higher latitudes. But the increase in some areas is not matching the decrease in others due to soil erosion, desertification, rising sea levels, and more.

Incredible changes in our food industry are on the way. We are transitioning from inefficiently extracting nutrients from grown animals to simply creating the proteins and nutrients for ourselves. New foods and food additives that complement our diets without all the emissions stand to disrupt longstanding agricultural traditions. And not just because it's more "sustainable" or "ethical" to ferment our own dairy or eat burger alternatives, but because animals are a particularly inefficient way of producing the end products we seek. More efficient products will simply be cheaper. Producing cultured meat will cost less than managing a farm full of cows from birth to burger. The many acres of forest now being clear-cut for animal grazing lands will be far less necessary for food suppliers going forward.

Precision fermentation can already produce so many different chemical products that getting them from animals in the future will simply make no sense. Nor will refining these chemicals from select plants, given the time and work necessary for cultivating and then processing plants. Plenty of the molecules we want—among them valencene (the citrus smell in oranges), vanilla, and even cannabinoids—are more cheaply produced by micro-organisms than from planting, harvesting, and processing plant materials.

Disruptions in ag-tech industries imply that longstanding historical exporters may no longer reign supreme for the products they grow and harvest. Countries with vast farmlands like Canada, the US, Brazil, and others in the EU may see local production of many products ramp up as they are suddenly geography nonspecific. And as with energy, meat and agricultural products may suddenly be far less centralized.

An artist's rendering of the Oceanix "floating city" proposal. See page 165.
CREDIT: OCEANIX/BIG-BJARKE INGELS GROUP

Another rendering of the Oceanix "floating city" proposal.
CREDIT: OCEANIX/BIG-BJARKE INGELS GROUP

DroneSeed operators piloting an ecorestoration and afforestation drone. See page 190. CREDIT: DRONESEED

Hypergiant's algae bioreactor. See page 201.
CREDIT: HYPERGIANT INDUSTRIES

Artist rendering of Carbon Engineering's direct air capture fan systems.
See page 205. CREDIT: CARBON ENGINEERING LTD.

CHAPTER 8

SMART HOMES, SMARTER CITIES

■

"What's bad about sprawl is not its uniformity,
but that it is so uniformly bad."

—JAMES HOWARD KUNSTLER, *HOME FROM NOWHERE*

In early 2019, the United Nations Human Settlement Programme (UN-Habitat) fielded proposals for futuristic sustainable city designs from research teams and architectural firms around the world. The plan was to develop realistic new methods of integrating urban renewable energy inputs, infrastructure designs capable of accommodating our twenty-first-century lifestyles in a post-carbon world, and strategies of buffering tomorrow's cities from the impending damages of climate change. That would include, for example, the imminent threat of rising seas for coastal communities.

Among the proposals submitted for consideration was a "floating matrix of miniature cities" design forwarded by a collaborative group involving the Bjarke Ingels Group (BIG), nonprofit Oceanix, and MIT's own Center for Ocean Engineering. The group essentially proposed a series of buoyant community pods fastened securely together, capable of supporting up to 10,000 citizens per city. The floating, modular neighborhoods would be energy self-sufficient—powered by sun, wind, and tidal energy sources. Cities would be freshwater autonomous with on-site advanced desalination facilities, rich with aeroponic plant-based food and aquaponic cultures, equipped to utilize zero-waste facilities, shared mobility systems for getting around, and—the entire point behind the proposal—each city would be largely immune to the threats of tsunamis, hurricanes, and rising seas. Island schematics available on the Oceanix website feature breath-taking views of high-tech conveyance systems, autonomous shuttles bustling people around town, massive solar, off-shore wind, and tidal energy farms—even some of those futuristic community domes you see in sci-fi movies (very much like the ones you see on the cover of this book). You will also find images of the floating city proposal included on page 160 in the photo section before this chapter.[1]

This is what many of us picture when we think about "smart cities" of the future. Chic, tech-savvy, highly automated, self-sufficient, and beautifully green. Probably a few elevated lanes for flying cars included just for good measure. But while ambitious city planners are certainly aiming for these lofty goals (floating car lanes aside), we'll first need to explore the many upgrades currently being implemented in today's cities right now. They may not be the "smart city" stuff of futuristic sci-fi housing project brochures, but they're certainly getting smart*er*.

Up to this point in the book, subjects related to "smart homes" and "smart cities" have been addressed alongside the specific renewable energy techs involved. Technologies like newer photovoltaic (PV) panels and solar windows were covered in the section for solar energy because they harness power from sunlight. But there are also new technologies for decarbonizing our modern lives that fit most aptly within this section, specific to improving the energy independence and resilience of our homes and cities.

If we want our societies to approach net-zero carbon emission status, their many features must require less energy. They must integrate more sources of renewable energy systemically. These are largely the needs of developing smarter homes for ourselves, as well: cut energy and resource expenditure needs significantly to become more sustainable.

Consider that powering residential and commercial buildings contributes a whopping 11 percent of total greenhouse gas emissions around the US.[2] An oversized component of those emissions is the energy we use to heat and cool our homes, here in North America. In fact, merely heating and cooling our homes generates approximately 441 million tons of carbon dioxide annually.[3] As a result, simply making our buildings more efficient can contribute toward dual goals of sustainable living and cutting emissions.

What kinds of new technologies can help reduce our home energy needs? Generally, insulation to save on heating; AC and shading developments to save on cooling; renewably produced hot water; smart grid-capable appliances; and anything else that will reduce our draw on energy resources.

Now, new insulating materials do not necessarily make for sexy book content. And many are not all that futuristic: techniques like advanced framing, high-performance walls, closed-cell spray foam, rigid foam boards, flooring insulation, and minimizing thermal bridging have gotten us pretty close to the insulation meeting net-zero building standards. But newer materials like nanostructured aerogels are becoming more standard for new homes, vastly reducing heat transfer beyond walls and through the roof.

Newly developed MIT aerogels are actually transparent and applicable for smarter *glass* insulation, providing another opportunity for advanced insulation (and cost savings). Windows outfitted with greater insulating material will be helpful for colder climates, allowing more sunlight to enter but not exit, like a greenhouse effect. Scientists also envision passive solar collectors with transparent insulation heating hot water pipes or other areas where heat is required all the time, rather than merely lining specific room windows.[4]

You will recall from the chapter on solar energy that windows have already received a considerable upgrade with energy harvesting tech. Transparent solar cell (TSC) panels have been used in greenhouse roofing and building windows, creating an opportunity for homeowners of the future to generate solar energy from window space. But incredibly, that means new smart glass windows will also have the ability to block selective wavelengths of light from even entering buildings. Transparent solar-harvesting systems will use molecular designs capable of absorbing only wavelengths of light that we cannot see—like ultraviolet and near-infrared—converting just those wavelengths into energy while allowing us to see the rest of our visible spectrum. Aside from generating energy, this will allow windows of the future to block out specific wavelengths of light to also bypass some of the heat our homes would otherwise be absorbing, and to potentially bypass the need for blinds on sunny days with electrochromatic shading control.

"Chromism" refers to chemical processes that induce a change in color, often with the connotation that the change is reversible. The term electrochromatic, then, refers to an electricity-induced chemical change in color. Within the context of windows, scientists have found that by putting materials with chromatic qualities inside panes of glass, they can alter a number of features with a very small electric input. The result is that smart windows allow for reversible changes in opacity and transparency. And by adjusting the miniscule amounts of voltage fed to nanoscale metal oxides, scientists have found they can induce opacity and reverse the effect thereafter. This means an ability to essentially dim window transparency with the push of a button (or an app, no doubt). So-called "dynamic daylight control systems" will curate sunlight more actively with light-controlling materials to control glare, dim light, or even redirect it evenly throughout a room.

Glass with opacity alteration means a new system of window blinds. The smart glass "TruTint" from supplier Nodis reportedly allows for instantaneous change while achieving up to 10 times lower costs in heat efficiency. The product offers an "infinite" number of tints and infrared control to optimize indoor climate and energy savings, according to suppliers. (Though I suspect the range of options is not literally infinite.) Newer developments allow for altered reflectivity in response to specific wavelengths of light, implying self-tinting windows for bright days.

Project Drawdown has estimated that adopting a technology like smart glass into largescale commercial buildings and in residential applications can result in a greenhouse gas reduction of 2.2 gigatons.[5] And the technology is particularly useful if it also saves us money. According to Drawdown, tests of electrochromatic glass in Japan have reduced cooling loads by more than 30 percent on hot days. And while purchase and installation costs may be twice as expensive as normal windows, energy savings make them more than competitive. Some newer models of smart glass may have the ability to harness some energy from the sunlight hitting your windows through TSC tech, resulting in even more savings.[6]

Some newer smart glass models will also be self-cleaning. Research indicates that applying a thin layer of titanium oxide to smart glass will degrade organic matter when exposed to UV light, making it less likely for dirt and debris to stay intact on windows for long.[7]

For homeowners who don't want to draw their shades via app, some new smart window models simply use coatings that offer similar—but offline—features like auto-shading. Smart window coatings are essentially filled with very small, water-filled balls that shrink or expand with temperature. Then, suddenly fibrous and increasingly heat-reflective, the spheres appear more like bundles of translucent material. One academic team published in *Joule* revealed that their film of microscopic particles alone could reduce temperature (without a loss of visibility) by as much as nearly 10° F.[8] The coat reportedly reflects up to 70 percent of the sun's heat away while letting in visible light, freeing up more of our energy from use on air conditioning.

Another team spun-off from MIT has produced AeroShield, a product with high-clarity silica aerogel placed between two panes of glass similar in appearance to a normal double-pane window. AeroShield's glass offers 50 percent more insulating power than normal windows (a benefit for colder months) with a payback period up to five times faster than other specially insulating windows. Or perhaps better yet: some researchers have been hard at work producing a DIY paint-on coating version of the tech, offering homeowners an energy-efficient window coating at one-tenth the cost of professionally installed retrofits.[9]

We may even be able to capture the thermal energy of incoming sunlight streaming through our windows, and—rather than harnessing the electricity through TSC—use materials similar to the thermal energy "hybrid" panels to *retain* it. To recap, thermal hybrid panels will soon be capable of placement on rooftops as traditional solar panels are, but they would capture thermal energy for either immediate or (through storage) later use. The device may bridge our much-needed gap in solar+storage technology. In terms of window tech, one new film with a uniquely designed molecular chemistry now offers an ability to capture thermal energy from solar rays and distribute it evenly throughout the glass. Materials have a yellowish tint at the start of the day, isomerize and turn transparent in response to solar heat, then reverse over time at night to re-start the cycle.[10]

As long as the sun is shining on the window's film, less heat can penetrate into rooms. But the added bonus is that heat can be stored for much longer periods than just hours, or even days. Researchers envisioned that by using a molecular solar thermal (MOST) storage system, homeowners may be able to store thermal energy for weeks or even months.[11]

The concept of "passive" buildings that trap incoming energy is now dominating smart home research. These buildings use as little heat as possible, reserve the sun's energy and heat, capture produced heat and recycle it all back into the home or save it for later, etcetera. In Canada, some smart home innovation companies have specialized in solar heat and heat-flow recycling to save in the winter. One sports arena in Colonsay, Saskatchewan, for example, heats its waiting and observation areas with waste heat captured from ice-making at its rink.

As for cooling homes on hot days, consider the feedback loop mentioned in chapter two: hotter days will increase our desire to use more air conditioning, producing more emissions, making days even hotter as a result. People in hotter, more populated countries would desperately like to install more air conditioning already. Developing new air-conditioner technology is therefore extremely important work. Smarter windows reflecting more thermal energy away from building interiors will help, but not nearly enough.

One of the most fascinating new developments in refrigerant technology involves the concept of radiative cooling. In the chapter about unique forms of clean energy, radiative cooling was discussed as yet another interesting take on generating energy from the world around us. But given that the concept implies a sky-facing surface should be cooler than the air around it, does that not also imply a plausible connection for our cooling needs?

Indeed it does. Because heat escapes along a thermal gradient leading out into the cold of space from the surface(s) of our planet, this has led some scientists to conclude that we might build materials to recycle the slight dip in surface-level temperature. And by setting up a specially structured reflective material, scientists were recently able to direct a wavelength of light back out into space that escapes our atmospheric cover more easily—leading to a radiative cooling effect even during the day. As a result, a material able to direct unique wavelengths more readily can become about 5°C cooler than ambient air temperatures.[12]

As you'll have guessed, researchers have now commercialized the effect. Burlingame, California-based SkyCool Systems utilizes small pipes filled with liquid just below the device to cycle colder water into buildings and refrigerant systems. Which means, with a 5°C (9°F) drop in cycled temperature, condenser units within conventional air conditioners could be replaced or upgraded. An intriguing trial run in Las Vegas saw one two-story office building reduce its electricity demands for cooling by 21 percent during the summer.[13]

A similar device was also developed by scientists at the University of Buffalo. It requires no added electricity to function. The system is essentially a low-cost polymer/aluminum film installed inside a box located at the bottom of a solar "shelter." The film performs radiative cooling and keeps ambient air within the shelter cool by re-directing more of the surrounding heat—with a sky-facing design that actively channels thermal radiation more directly from the film into the atmosphere. The device helps to corral and focus directed heat away from ambient surroundings without any use of electricity. According to the team's research, temperature reductions around the enclosed space dropped by about 6°C (11°F) during the day and 11°C (about 20°F) at night.[14]

You might be surprised to learn that refrigerant management actually topped Project Drawdown's Top 100 solutions for climate change. This is because when refrigerant materials are managed poorly, they can be devastating to our climate.

For decades, our refrigerators relied on a class of chemical compounds called chlorofluorocarbons (CFCs) that absorbed heat and then released it somewhere else. When we realized that CFCs like freon were damaging our ozone layer, we actively phased them out. New refrigerators, however, have relied on similar chemical compounds, called hydrofluorocarbons (HFCs). And we seem to have created a similar problem with HFCs in that they still reportedly warm the atmosphere more intensely than carbon dioxide—at a scale of between 1,000 and 3,000 times worse. In October of 2016, however, more than 170 countries convened in Rwanda to roll out mutual agreements to phase out these chemicals.

One new tech for cooling without toxic chemicals involves the process of cycling heat throughout larger networks of ambient systems. So, while it doesn't help smaller users, larger AC culprits like data centers, labs, and hospitals can benefit. Durham, North Carolina-based Phononic specializes in producing a chip to drive heat within a network's entire shell, creating an environment in which there is never any singular "hot spot."[15] This offers progress in reducing toxic HFCs.

Another way to endure new levels of heat without air conditioning is to try reflecting more sunlight back into space before Earth's many surfaces can absorb it. This is an option we will return to again in a later chapter while covering new "geoengineering" initiatives that aspire to engineer our planet in very direct ways to off-set climate change. One geoengineering initiative falling under a category of solar radiation management (SRM) includes painting our rooftops colors that reflect away more sunlight to reduce the amount of heat absorbed. This is certainly an option to consider for reducing citywide temperatures.

So-called "cool roof" coatings can not only reduce the amount of heat warming up a home, but also slightly reduce the surrounding air temperature—particularly if many homes in an area use them. New research from the Department of Energy's Lawrence Berkeley National Laboratory has shown that widespread adoption of cool roofs can also lower overnight temperatures during hotter evenings as well.[16]

Cool roof materials will definitely benefit from new applications in nano-scale surface material developments. For example, a growing number of nanophotonic materials have been developed, allowing scientists to manipulate the way light encounters materials at the nano scale. One new development in materials sciences from Berkeley-based Cypris Materials involves self-assembling reflective coatings offering tailored optical properties for even more selective wavelength reflection. The company's coatings enable more heat mitigation by reflecting ultraviolet and near-infrared radiation like the models above—but rather than merely applying to windows, Cypris's coatings could be also placed on walls and rooftops to reflect more of the sun's rays with a transparent, colorless coating. Cypris's technique for doing so involves a more specific bottom-up approach in scaffolding unique polymers to self-assemble into nanostructures (and replacing multimillion-dollar fabricating equipment costs in the process).[17]

There are also promising new tech upgrades for smarter cities and buildings that aren't quite as flashy as nano-coatings and radiative cooling panels.

Consider the concept of "district heating" factoring into future city planning. As it turns out, the energy needed to provide heat for several buildings can be less than the energy equivalent of buildings heating themselves individually. The trial community of Nordhavn in Denmark is a case study in using this phenomenon to save energy on larger scales. The area utilizes community-level heating and cooling, allowing for massive energy savings in shared inputs. The district is smart grid integrated, allowing for an unprecedented real-time awareness of how heat and electricity are being applied in separate buildings and even separate appliances.

Further, the inputs and outputs can all be synced to a smart, flexible operations system that maintains both an electrical grid and another new concept here: the thermal grid. A thermal grid can take advantage of similar AI and machine learning inputs to maximize heating efficiencies, as smart grids will smooth out peaks and valleys of energy use. A good thermal grid will recycle waste heat from locations like power plants and data centers, for example. And given that Nordhavn is a coastal community, circulating cold seawater throughout the district's cooling system infrastructure has allowed engineers to reduce up to 80,000 tons of CO_2 amidst energy savings.[18]

New advancements in internal room sensors and smart thermostats can also assist us with our heating and cooling issues. Intelligent thermostats can now access weather forecasts from news sources and plan internal temperatures accordingly. Smart thermostats can even communicate with the grid at large, assessing energy intake to match "low-carbon electricity."

In terms of saving on resources in homes (simultaneously saving the city at large from shuttling them in), nothing may be more upending than the ability to procure one's own water. And now, tech similar to the devices used by scientists to draw in moisture from ambient air and harness the molecular portions of water can also be used by homeowners—in the form of solar panels able to draw in and sequester water for domestic use.

Tempe, Arizona-based Zero Mass Water produces a line of solar "Hydropanels" that not only provide renewable energy off the grid, but also reportedly produce between 2 to 5 liters of drinking water per panel, per day (relative to sunlight and the surrounding humidity, of course—about a half-gallon to 1.3 gallons in conversion). In terms of a commercial equivalent, that is roughly 10 average (500 ml, 16.9 fluid oz) water bottles per day, with an option to scale this effect into larger field arrays for a cumulative effect. Essentially, these Hydropanels use solar power to pull in large amounts of air, then collect moisture from that air onto an absorbing material with solar heat converting it into liquid water. The device then channels that water into a 30-liter (7.9 gallon) reservoir where it is mineralized and kept clean using ozonation, or into a large tank for commercial-scale needs, where the same process occurs. Incredibly, reservoirs can also be run directly to internal faucets. Engineers with Zero Mass have more recently unveiled an additional series of system sensors to fine-tune owners' monitoring of the water reservoir itself. Sensors will allow customers to keep tabs on the water quality in real time and monitor daily reports via the Zero Mass Water app.

Or consider one of my favorite new water-procuring tech developments: next-gen, ground-level dehumidifiers powered by renewable energy.

Researchers at UC Berkeley have developed a water harvester capable of pulling more than five cups of water from low humidity air per day. The device collects water from the atmosphere in very dry conditions and then releases it for harvesting as liquid drinking water. Rather like the dehumidifier in your basement, which absorbs airborne moisture and fills a bucket underneath, this device works at low humidity and in dry conditions.

The metal organic framework (MOF) used by UCB scientists has increasingly become a centerpiece of new technologies, given that powdered materials full of organic molecules mixed with metals is relatively new for us at large scales. Water molecules simply floating around in ambient air become affixed to the MOF somewhat like a sponge, and water is later extracted into a tank where heat creates droplets that form a collectible basin of pure H_2O. Researchers' start-up company Water Harvest Inc. will reportedly be making a microwave-sized water harvester able to produce water even in the desert, with a second refrigerator-sized unit able to provide water for a household of four.[19]

Venice Beach, California-based company Skywater (or Skysource Alliance, as they have also called themselves) has also created a new device for harvesting atmospheric water from the air around us. Skywater/Skysource has notably won the Water Abundance XPRIZE for developing a renewably powered and scalable water harvesting device that generates drinkable water through a patented Adiabatic Distillation Process. Water vapor is essentially again gathered from the air and then turned to a liquid through the team's equipment. The result, in the words of the XPRIZE team judging the device, is "an easily deployable high-volume water generator that can be used in any climate, meeting the competition parameters of extracting a minimum of 2,000 liters of water per day from the atmosphere using 100 percent renewable energy, at a cost of no more than two cents per liter."[20] Two thousand liters, for reference, is about 528 gallons. And at 2 cents per liter, their technology may be a game-changer in disaster zones and poor, rural areas.

Skywater's technical details describe how airborne water vapor can be captured through an air filter. A condenser heats and subsequently cools the vapors in air to maintain the "dew point" at which vaporous water turns to liquid. That water is then collected, purified, and filtered to produce drinkable water. The device is then also rigged to what the team calls WEDEW (or "wood-to-energy deployed water"), a biomass gasifier system allowing for fuel in the form of woodchips—or any other mulched biomass—to power the condenser system, although they say the system could also run on solar power or a battery backup. A small unit designed for office-level use reportedly costs around $1,500. However, the team also believes they can scale that price down to about the cost of a new, household AC unit.[21]

Durban, South Africa-based Water from Air has created a device somewhat similar to Skywater/Skysource. Air with water vapors passed through a filter and a compressor produces liquid water, which a UV light and a subsequent carbon filter then purify to produce clean, drinkable water from the surrounding air. The water then undergoes a reverse osmosis process through another filter to add minerals, before passing through a second UV light sterilization stage.[22] What you end up with basically looks like a larger water cooler, around which office colleagues or family members could gather to discuss how cool it is that engineers can now obtain water from air.

Meanwhile, Virginia Tech University has created a device that resembles a harp which is designed to harvest the water droplets in fog.[23] Water droplets carried by the wind within fog pass through the device, which allows droplets to gather along points on each vertical wire until they gain enough weight to drop down into the collector tank below. The team at Virginia Tech calls their moisture-catcher a "fog harp" and claims that it is a threefold improvement over the fog nets currently being used in arid and semi-arid climates. The result is an efficiency of about 2.4 gallons of water gained per 10 ft² of material. The team was reportedly inspired by plants that use their own variation of fog harvesting in a similar fashion, like redwood trees on the California coast that have arrays of branches and leaves running parallel to each other.[24]

Of course, every new development for smart homes can scale to support large, commercial buildings. Advanced new "passive" high-rise buildings—in which energy consumption has been slashed by 60 to 70 percent—are already being used around the world.[25] Thermostats often range only from 68 to 75°F to reduce any extremes. Thick, insulated walls and triple-paned windows significantly minimize the need for extensive heating and cooling systems. Some residential towers even display a digital tally in the lobby that tracks energy consumption by floor, encouraging residents to do what they can. One such high-rise, The House on Roosevelt Island, projects annual greenhouse gas savings of around 882 tonnes of carbon dioxide with passive heating and cooling tech.[26]

The Brock Environmental Center in Virginia Beach, Virginia, may be a model for future smart buildings. The facility reportedly produces all its drinking water from collected rainfall (the first commercial building in the US allowed to do so). It therefore uses 90 percent less water than commercial buildings of similar size. Moreover, distributed energy systems linked to the building allow it to produce 83 percent more energy than it consumes.[27]

Increasingly self-sufficient buildings are now becoming more commonplace in absorbing enough energy to get by independently while also harvesting enough raindrops to yield a net-zero water balance for all needs. The Bullitt Center in Seattle, for example, is a six-story commercial office building in the downtown area achieving both feats. Rainwater is collected in a cistern in the building's basement, filtered and then passed through a UV disinfection system before practical use in low-flow fixtures in sinks, showers, and kitchen areas. The cistern is capable of holding about 52,000 gallons, offering roughly a month's worth of water. Meanwhile, the water that Bullitt Center uses for its foam-flush toilets is filtered but not disinfected. The low-flow fixtures help to conserve significant amounts of water and the foam-flush toilets only use approximately two tablespoons of water per flush.[28]

Standardized assessment protocols also now exist for gauging the emissions standards for various materials throughout construction. This is especially encouraging because of how complicated it must be. Any one construction project may contain literally thousands of different materials, all with varying carbon emission impacts.

Immense tracking applications for materials worldwide have been entered into an Environmental Product Declaration (EPD) increasingly utilized by more construction companies to chart carbon cost for different materials. Anything from insulation and ceiling tiles to metal framing and insulation materials are included. New tools like EC3 (Embodied Carbon in Construction Calculator) use Microsoft-driven analysis of thousands of EDP items to calculate carbon impacts more quickly. We will explore the concept of "regenerative architecture" as one strategy among many others to sequester more CO_2 over the coming years, but for now, yet another avenue exists primarily for cutting emissions while still producing first-class buildings. And there are a variety of motivations for companies using these tools, aside from seeking net-zero building status. For example, companies may wish to know which materials are cheaper in the event they are someday be taxed for carbon emissions.

SMARTER CITIES COMPOUNDING
SMARTER EVERYTHING

As we have explored new technologies throughout this book, we have appropriately skipped some of the newest tech upgrades that do not directly apply to energy savings or re-balancing our climate. But a variety of incredible new scientific developments *do* exist that have furthered the form and function of our cities aside from these goals.

Consider, for example, the advancements in artificial intelligence and machine learning discussed earlier. Couple those applications with the latest developments in our increasing internet-of-things connectivity and the deployment of self-powering sensors, capable of detecting various system changes that we might want to measure. These tools are not only available to homeowners: smart city operators will increasingly utilize these tools to make our cities more livable.

When a domestic terrorist attacked multiple people in a targeted series of stabbings in New York in 2019, the suspect—who had fled the scene before police arrived—was apprehended only an hour later and swiftly taken into custody. How? Camera systems deployed throughout New York City remotely detected the license plate on his car before routing that information immediately back to the city's crime lab for escalation. A remote sensor system using advanced vision analysis had captured the unique search-string of his plate number, assessed the visual cues for relevance with AI-assistance, and then patched that information along to authorities, who then knew his car's exact location. Tools like these will make our cities safer and more functional in new and interesting ways (although hopefully not matching these tech upgrades with facial recognition and subsequent search-and-track software systems, as China increasingly has used).

Cities of the future will take advantage of solar storage systems on lights, kinetic energy sidewalks, and more. But perhaps no other tech upgrade will benefit cities more than internet-of-things connectivity. More connectivity between city-wide energy systems, service providers, and traffic operators will yield some interesting new outcomes.

Smart cities (and increasingly smart municipalities) will also benefit from the IoT application of shared data of connected vehicle fleets—not just the connectivity of all public vehicles sharing their data, but the shared data of *all* cars on the road. Consider the incredible series of data points that vehicles can potentially share: ignition on/off data, accelerometer, GPS location, analysis of road data from autonomous vehicles, and more. Even without sharing VIN/ID numbers, all of these data points can be assembled and analyzed in real time on advanced networks—and then processed by machine-learning algorithms to produce desirable outcomes for all drivers.

Compiled information can be assessed to produce meaningful data for city operators. For example, the combined vehicle data on hard braking, unusually fast speed reductions, and airbag deployment can instantaneously alert various first responders of accidents. Or consider traffic jams. Sure, using an app like Google's Maps network can alert drivers of congestion ahead, but autonomous vehicles feeding into a central data network can know much more about road conditions, re-directing for alternative routes.

A company like Los Angeles-based Waycare already utilizes machine learning algorithms to generate traffic management solutions for cities and commercial fleet owners. The result has reportedly been beneficial for reducing congestion and increasing accident response times, making this kind of technology undoubtedly a benefit for city professionals. In fact, the Nevada Highway Patrol ran a pilot program using insights from Waycare's network. The results of real-time information sharing were a 17 percent reduction in crashes along portions of the highway in question and accidents identified up to 12 minutes faster.[29] Companies specializing in these services or smart cities running their own programs can accumulate immense amounts of data points to improve roadways, like historical data of congestion and accidents. And analyses of historical data can lead to predictive recommendations, saving on future congestion and accidents.

Smarter cities will also process and treat waste more effectively. I have reserved some aspects of smarter cities like improved waste management and wastewater remediation for a later chapter on damage control, but it's worth noting in advance that even the basic structure of how these dynamics operate can be improved on a city-wide level.

An incredible amount of energy goes into shuttling waste materials and water from our homes to a treatment plant. Every drop of waste liquid leaving our sink drains, our shower drains, and down the toilet pipes has to be transported underground. From there, it all needs to be transported to the outskirts of town where it is treated and then discharged. And in a world where the average person flushes 4 to 8 times per day and often enjoys showering once a day, transporting, treating, and returning recycled water to us is becoming increasingly unsustainable. What can we do?

Consider the aforementioned researchers and new companies utilizing bio-based remedies for wastewater. In some cases, researchers have managed to produce biofuel, other times specialty chemicals or fertilizers. These capabilities now provide us the means not only to do something sustainable with wastewater, but also potentially provide societal bonuses from it.

One solution to our wastewater needs seems to be new innovations in creating more localized, community-level wastewater treatment plants. The twist? Some companies are turning them into chic, urban greenhouses with diverse sets of flora inside—all supported by various levels of organic material created by waste-consuming bacteria. Princeton-based Organica Water is doing precisely this with installations in Hungary, China, India, and Indonesia serving as case models (on top of another 80 installations worldwide). Organica's approach reduces energy expenditure because a localized destination for wastewater reduces transport costs. Utilizing the help of engineered bacteria cleaning it essentially for free does so further. Powering such a facility with renewable energy helps build sustainability further. Providing urban biomass in greenhouse fashion is another added bonus.

Normally, nobody wants to live near a wastewater treatment plant. But when it's a state-of-the-art urban greenhouse? Certainly less of an issue.

Speaking of greenhouses, there are also an increasing number of green spaces factoring into smarter cities. New innovations on integrating more greenery into city life have focused on green roof installations and the incredible new "vertical forests" of buildings.

Vertical forests are essentially high-rise buildings draped in a living plant canopy, integrated by cities to introduce more trees. I recommend searching the internet for the "living buildings" already in operation, courtesy of Italian architect Stefano Boeri, creator of Milan's Bosco Verticale (read: vertical forest). Boeri's latest forest city innovation in Cancun, Mexico, will reportedly feature 400 different species of plants and trees—including roughly 7,500,000 plants in total and 200,000 trees altogether. The planned "human to tree" ratio is roughly 2.3 trees for every city inhabitant. The aim will be to achieve total energy and food self-sufficiency when the project is complete.[30]

Or consider the existing vertical forests in the Dutch city of Utrecht (construction set to finish in 2022). Or China's Nanjing. All of them basically feature roofing, awnings, and balconies covered by vines and shrubs. Several also feature roof-water retention systems to provide nutrients to the attached plants.

One upcoming project, currently in technical study stages, is plotting out project lands just north of the Chinese city of Liuzhou. The plan, if successful, will house 40,000 trees and nearly one million plants spanning more than 100 species distributed throughout building walls, parks, and street-level implementation. The effort will absorb nearly 10,000 tonnes of CO_2 per year.[31] That is just a drop in our collective bucket, but may inspire other new building projects to follow suit.

Cities planning for twenty-first-century challenges like extreme storms and heavier floods are also increasingly using green spaces as mitigation areas for floodwaters. For example, newer infrastructure designs in smart cities include strategically placed rain gardens (which collect water even as they augment the trees intending to reduce the heat island effect), installing permeable pavement with collector channels below, and "bioswales"—essentially gaps between paved areas with soil and trees capable of catching some of the excess water flowing by. Those soils can also be equipped with filtration layers (read: bioretention strips) to capture more of the pollutants from storm waters far before they reach adjacent rivers and affect marine life.

Research also indicates that lining more neighborhoods with trees, initiating green rooftop programs, and installing vertical gardens along traditional infrastructure can present a host of benefits for urbanites. The reasons for green spaces are now manifold: plant life generates oxygen, absorbs CO_2, creates a positive aesthetic, and reportedly lowers the "heat island" effect of raised temperatures in urban areas. In fact, the US Environmental Protection Agency has estimated that lining a suburban neighborhood with trees can reduce temperatures in that area by about 4–6°F (2–3°C).[32]

For these reasons, we now see urban green spaces integrating tech and greenery together. A company like UK-based Green City Solutions markets moss-lined and IoT-laden benches, bridge covers, and bus stop canopies for absorbing at least small amounts of CO_2 while integrating a regenerative biofilter for sponging up air pollution. And if it looks good, serves a purpose, and contributes to helping off-set some of our climate troubles, I suspect we'll likely see it featured in cities of the future.

Finally, it's often assumed that futuristic smart cities will have plentiful, drinkable water. Scientists have been focused on improving this concept, given the importance of water for human life. And while water is a basic human necessity, our sources of fresh water may actually decrease as time goes by in our changing climate. Aquifers have dried in hotter regions, ice pack does not supply major reservoirs with nearly as much water each year, and more. Fresh water already makes up only a small fraction of the water on our planet and predicted shortages could make life far more difficult for those living in at-risk regions.

Water harvesting technologies on the level of individual families and communities are progressing rapidly. Atmospheric water harvesting tech was covered earlier in this chapter, along with solar panels capable of producing impressive amounts of water for residential use. But incredible new advancements have also taken place in the area of desalinating seawater, at both community-level and industrial scales. New work in desalination includes a variety of new upgrades ranging from nano-scale water filtration and ionic membranes that literally pull the salt from water, to next-gen solar stills and solvent separation techniques.

One new device developed by scientists at the King Abdullah University of Science and Technology in Saudi Arabia has mixed PV panels and solar stills—but unlike the tech from Zero Mass Water, uses the waste heat from panels to power the distillation process. The device uses heat that would normally just reflect off PV panels to vaporize channeled seawater below, capturing the subsequent water vapors through a porous membrane. The distilled water is then directed to a collector component adjacent, all without compromising the PV panel's energy output. The device can reportedly produce about seven cups of water per square meter of the panel's surface (roughly 10 square feet) every hour.[33] That is roughly double the water output of conventional solar stills.

Another team at the University of Texas has repurposed nature's signature rose petal design to more evenly distribute water as it is being steam-purified, evaporating salt and other impurities in the process. Using materials that cost less than two cents, the team was able to develop a device that produces roughly a half-gallon of purified water for every 10 square feet of coverage. Their device resembles a black rose in a jar—which, unlike most current solar "steaming" techniques for water purification, is not bulky and expensive. They also used polypyrrole, known for having advanced photothermal properties in converting more solar light into thermal heat (which then helps drive the purifying process for this application).[34]

One team from the University of Bath and the University of Johannesburg has utilized 3D-printable materials coupled with solar energy to power a mobile, quick-setup ionic separation device to purify water in disaster-stricken areas. The team's device employs solar power to move salt anions (negatively charged ions) and salt cations (positively charged ions) between chambers through a micro-hole membrane. With a one-way flow design, the device essentially pumps salt through the membrane rather than water through a filter.[35] Only about 50 percent of the salt in seawater could be removed—taking out 90 percent is considered the goal—but this novel use of 3D-printed, mobile and "ionics" basis for purifying water could potentially be a game-changing combination when developed.

Scientists at Columbia University have also helped to improve desalination by developing a method for treating much of the brine produced by traditional methods. Brine is the extremely salty liquid left over following desalination, which then must be repurposed or disposed of. On average, desalination services across the globe pump out 142 billion liters of brine (roughly 37.5 billion gallons) every day.[36]

The team at Columbia utilized a technique called Temperature Swing Solvent Extraction (TSSE)—often used in chemical engineering and purification of industrial materials—to chemically separate seawater from its salt content. Incredibly, the technique was capable of treating brines up to seven times as salty as seawater, a marked improvement over current techniques only capable of handling brines about twice as salty. Moreover, the team was able to remove approximately 98.4 percent of salt from water, comparable to our current gold standard of reverse osmosis. Except, unlike reverse osmosis, the process was less expensive and able to treat the hypersaline brines that osmosis cannot. The team believes that solvents may potentially disrupt an industry currently led by advanced membranes, high heats, and evaporative phase-change techniques.[37]

Ultimately, "smart cities" will continue to evolve and integrate a variety of new features. Increasing electric charging infrastructure throughout urban spaces, encouraging zero-carbon modes of transportation, retrofitting inefficient buildings, and setting expectations for new buildings to be sustainable will all improve urban life. We will undoubtedly see a vast expansion of distributed energy sources around our cities supporting the increasingly regional, independent, and smarter grid systems being developed. We will see more green spaces. More community-level district heating and cooling, such as Nordhavn. It can't all be futuristic domes and flying car lanes—but advanced enough that our cities are destined to be smarter overall.

CHAPTER 9

STEPPING BACK FROM THE BRINK: REVERSING CLIMATE CHANGE

■

It's worth pointing out there is no scientific support for inevitable doom. Climate change is not pass-fail. There is a real continuum of futures, a continuum of possibilities.

—KATE MARVEL, ASSOCIATE RESEARCH SCIENTIST, NASA GODDARD INSTITUTE FOR SPACE STUDIES (2018)[1]

Despite all the good news that we have explored throughout this book so far, it's now time for a necessary gut check. It is entirely appropriate that we allow ourselves to feel encouraged and inspired by the incredible progress in clean energy and sustainable living under development around the world. But we must also remind ourselves that even if we do transition away from fossil fuels toward a zero-carbon society—even if we transition toward a zero-carbon *world*—it may still not be enough to save us from serious environmental damage.

Our atmosphere is full of greenhouse gases. In response, we are developing clean energy technology to prevent the production of more such gases. But the grim reality is that we still have an atmosphere full of greenhouse gases to deal with. The emissions we have already released over the decades will linger, warming our planet for years to come. Hence if greenhouse gases are driving destructive climate change and cutting our emissions will not reduce the amount of pre-existing gases quickly enough, any hope of reversing the destruction predicted for our next few decades will need to focus more directly on removing greenhouse gases from the atmosphere.

It is difficult for scientists to estimate exactly how long the additional warming will continue even after we reach a zero-emissions state. What scientists do know is that our current level of over 415 parts per million (ppm) carbon dioxide in the atmosphere is sufficient to cause permafrost thaw, which may trigger not just the incremental warming we're used to—but possibly a non-linear tipping point from which it will be very difficult to recover. And that, coupled with more uncertainty regarding when we might reach a zero-emissions state, necessitates the active removal of greenhouse gases as soon as possible.

This chapter will focus on how scientists intend to clean up our atmosphere. Their plan, generally speaking, will be to hasten a state of "drawdown" where greenhouse gases stop rising and start declining on a global scale. We will explore the strategies being developed to achieve not just carbon neutrality, but a *carbon negative* society in which we actively remove emissions from the air while emitting even less. We'll explore the strategies we have, how those strategies work, who is working on them, where they have already been deployed—and how it all contributes to our portfolio of new options for the future.

The latest reports from the Intergovernmental Panel on Climate Change (IPCC) now discuss the role of so-called "negative emission technologies" (NETs). These technologies, which aim to sequester carbon dioxide or recycle it, now seem expected by the IPCC in "most mitigation pathways" toward a more sustainable range of temperatures.[2]

NETs for removing carbon dioxide can ultimately range from the incredibly simple to the complex. They can be as straight-forward as planting more trees or as sci-fi imaginative as self-sustaining carbon scrubbers powered by sunlight and plucking CO_2 molecules right out of the air. And while researchers cannot often agree on which NET option will help us the most, the majority of models now agree that we'll probably need an assortment of them working in unison for a real shot at saving ourselves.

Realistic projections for our planet's future climate often contain the ominous warning that if we do not begin developing and deploying NET options soon, we will have to rely on them even more in the future. This is not a situation in which we want to find ourselves—threatened with more climate-induced suffering and increasingly reliant upon untested and potentially expensive technology. Deploying early test trials for these technologies is therefore crucially important.

It is equally important to remember that most academic assessments of NETs argue that they cannot encourage complacency on reaching net zero emissions. They cannot perpetuate "business as usual" or reduce our urgency in cutting emissions. In fact, by some estimates, our window for operating on a "business as usual" schedule will only give us roughly a few years left at our current emissions rate before the deployment of NETs will become essentially unavoidable.[3]

Timelines for NET deployment usually predict testing phases to ramp up over the next few years. We will then see headlines about the results and their implications sometime between 2021 and 2025. At that point, researchers should have a much better idea of how these technologies are impacting the actual environment. Rapid international deployments are then projected to follow roughly around the year 2030, with truly international deployments slated for the latter half of this century.[4] The goal, by some estimates, will be to scale up these techs quickly and reach at least a plateau of removing about 10 gigatons per year of CO_2 (10 Gt/yCO_2) globally by mid-century.[5]

Some NETs will require significant investments. Others will not. Each option comes with a suite of positives and negatives in terms of cost, land use, safety, resource management, and more. So while there is no "silver bullet" option, there may be, as some commentators now call it, some "silver buckshot"—a combination of options we can use.

The cheapest and easiest way of working toward drawdown is simply to plant more trees. Trees naturally absorb CO_2 from the surrounding air as a part of the photosynthesis process. A portion of that absorbed CO_2 is then converted into the biomass that we see as wood, leaves, and bark. Another portion is stored in the roots below. Afforestation (planting new trees) and reforestation (restoring forests we've lost), or just stopping deforestation, are simple, easy, and very "natural" options for helping to get the planet back to its preindustrial carbon cycle.

Crucially, there are differences between how much CO_2 is absorbed and stored by different plant and tree species. So besides simply planting more trees, we can plant an assortment of specific plants and crops offering us greater amounts of CO_2 storage. But even with some plant species offering us greater carbon capture, there is still a significant portion of carbon being returned to the atmosphere when plants decay. Therefore an even better strategy may be not only deploying crops with superior carbon capture, but also harvesting them before they die. Doing so would give us a chance to capture their stored carbon before it is released back into the air. We can take a selection of higher carbon-storing crops and burn them after harvest or use them as fuel feedstocks to gain their energy while capturing the CO_2 emissions from a more concentrated source. This option, called bio-energy with carbon capture and sequestration (BECCS), has received a considerable amount of focus among climate response strategies (undoubtedly because market forces will help drive the process).

In researching which plant species will offer us superior amounts of "carbon sink" in removing CO_2, we've found that some are in a class of their own. So-called "blue carbon" sites exist in mangroves and coastal wetlands, earning themselves the blue moniker because they grow partially submerged in water. Sometimes they are even found in salt water. These plant species not only store more carbon than terrestrial plant life, they also offer us a variety of other bonuses like coastal wildlife preservation and a buffer zone for the impact of hurricanes. They therefore offer us a natural and well-understood NET option for scientists to develop plans for deploying immediately.

On the less natural end of the NET spectrum, we have options like direct air capture (DAC). These are the carbon capture machines you've likely seen featured in mainstream media that look like giant fans and air purifiers. Air goes in one end of the purifier, a series of chemical scrubbing actions take place inside specialized filters, and air with less CO_2 goes out the other side. The captured CO_2 is then shuttled away for utilization or sequestered safely underground.

DAC options offer us both positives and negatives. Strategically placed filters that capture more CO_2 than plant life can save us incredible amounts of land and energy resources (if they are renewably powered)— particularly if we need to wait for trees to develop but can deploy DAC machines immediately. However, this option has also been prohibitively expensive to date.

Now, with just a few options and introductions out of the way, let's dive deeper into the possibilities that scientists are exploring—and the advanced new innovations making negative emissions technologies cheaper, safer, and more reliable over time.

ALL NATURAL AND NATURAL-ISH: AFFORESTATION, REFORESTATION, AND POTENTIALLY SUPERCHARGING PLANTS

Planting more trees . . . It seems so simple that it can't possibly be a solution to global, systemic climate change. We're up against worldwide, institutionalized, multinational corporate machines with fossil fuels. And yet, while the immediate, small-scale impacts of a few individual trees, crops, and forests can seem spurious, scaling our forestry efforts up to the size of an *entire country's worth* starts to produce CO_2 storage rates worth talking about.

The good news is that by some estimates, we still have enough room left in the world's currently existing parks, forests, and abandoned land to plant about 1.2 trillion trees,[6] which would clearly be a game-changer for remedying climate change. Of course, nobody really expects us to fill up all that land entirely with trees—but if we did, ecologist Thomas Crowther at ETH Zurich in Switzerland has estimated that it would cancel out about a decade's worth of emissions. Crowther was also involved in another study published in July 2019 estimating that if we planted just 500 billion of those trees, we could potentially remove about two-thirds of the CO_2 we have pumped into the atmosphere since the Industrial Revolution.[7]

There are some caveats, however. For one, planting trees is not an ideal scenario for some soils and regions. We must consider the land requirements for feeding global populations as well. Some local ecosystems may also be more adapted for grasslands and their associated wildlife. In fact, one study from scientists at the University of Cambridge (UK) found that river flow can be significantly reduced in areas where trees have been planted, to the point where some river flows may not recover.[8] There's also a chance that altering the planet's albedo with treetop cover might cancel out a portion of the drop in warming that we'd like to achieve. Moreover, planting incredible amounts of trees will only pay real dividends when they begin to mature—and that's assuming we do not consistently lose trees planted to burning or deforestation, or a disease in the event we're foolish enough to "monocrop" just one tree species.

There has also been plenty of academic criticism that papers like Crowther's have been too generous in their carbon counts.[9] This is especially so with new research telling us that carbon storage capabilities for some tree species will decline over time if the world's climate continues to change.[10] Although at the very least, experts can all agree that planting great numbers of new trees will help us—and that the best time to start is right now.

The take-away seems to be that planting trees is a low-cost, low-tech, all-natural option for storing carbon dioxide. We can implement the option immediately, and it will scale to make a meaningful impact against climate change. Scientists may argue about where to plant, how many we need, and what exactly the final impact will be, but at a minimum, planting many trees immediately can make a significant contribution to our buckshot.

Part of the problem with planting trees, however—aside from the land required—is the demand of labor in accomplishing the task. We need human capital with real volunteers going out to plant more trees. Unless private companies, NGOs, or governments start to bankroll more tree-planting initiatives, we will also need people to donate time and money to this cause. It's not always a proven method for achieving larges-cale, planetary change.

Fortunately, we have improved afforestation methods considerably over time. New developments in tree-planting technology have freed up humans to plant many more trees per day than we could before. Unmanned aerial vehicles (UAVs) have allowed us to automate large portions of the process—notably the most energy-intensive and sometimes dangerous portions. With the help of autonomous navigation tech, satellite imagery, and on-board AI machine-learning in seed-planting drones, we have gained an unparalleled ability to forest large areas of land in small amounts of time.

Anyone who has ever planted trees manually knows it's a grueling process. The photo-ops of politicians planting a single tree with smiling families do not do justice to the ruthless grind of a ten-hour day spent hauling around large sacks of seedlings and getting them into the ground. But now, drones flying by themselves or remotely by operators can survey the land to be seeded, position themselves overhead, and disperse germinated seed-pods in a variety of ways. And multiple new companies around the world are pioneering efforts to push this tech forward with immense land restoration projects.

Consider Oxford's start-up BioCarbon Engineering. BioCarbon begins their restoration projects by launching a surveillance drone to map terrains and develop a 3D plot of the area to be seeded. From this aerial survey, software can then extrapolate an algorithm to craft a unique series of planting sites for seed drops. With this data on board, then come the seed-firing drones flying to programmed locations, launching germinated seed pods into the ground. The process is several times faster than our human planters can manage, and it even opens up new planting sites normally off-limits to human personnel (like steep mountainsides). The company also offers drone-assisted grass restoration, flower-planting, bush sites, and more.

At this time of writing, BioCarbon has experimented with using drones to disperse more than 38 species of seeds across Australia, Myanmar, and the United Kingdom.[11] Working alongside researchers with the non-profit Worldview Impact, the company was once tasked with restoring roughly 350,000 hectares of coastal forest lands in Myanmar. Numbers like those could be a non-starter for a manual team's limited resources. But given BioCarbon drones' capabilities thus far, Worldview Impact estimates that two operators (each operating up to ten drones apiece) in different locations could theoretically plant hundreds of thousands of trees *per day*.[12]

One might argue that drones firing seed pods into soils can be disruptive for local ecosystems. Arguably, though, hands-on afforestation efforts with similar intentions might be even more invasive.

In the near future, drones able to navigate environments on their own could deploy germinated pods day and night. But for now, even available technology can conduct aerial reconnaissance and navigate environments largely on their own, inspect optimal planting sites, and spread seeds as they fly with only minimal supervision. Wireless communication can allow drone operators—potentially far removed from drone locations—to analyze the data being collected about an area's topography and soil conditions. They can then use that information along with satellite data to queue drop-zones for better seed dispersal going forward.

Seattle-based DroneSeed reports that their drones are able to propagate about 40 acres in a single day—roughly six times faster than our human planters. And unlike other seed-deploying drone companies, DroneSeed uses FAA heavy-lift certified drone swarms, each unit capable of hoisting 57 pounds of germinated seeds per flight. They're essentially the heavy gunners of seed deployment. DroneSeed utilizes a sophisticated imaging stack, first conducting a surveillance pass using waypoints from satellite imagery. They use lidar to collect spatial information at high resolutions to produce a similar 3D map of a terrain's vegetation. On-board systems can then identify unique features like stumps and roots, or even bushes and signs of erosion. From there, a swarm of drones (up to five in the air at a time) take flight to deploy nutrient-packed seed pods to mapped locations. So in this case and most others, the drones fly to pre-planned locations rather than using sight-based autonomous navigation. An image of DroneSeed pilots operating the equipment can be found on page 161 in the photo section right before Chapter 8.

While many drone-assisted foresting companies offer specific services like land restoration and wildfire recovery efforts, Sydney-based AirSeed Technologies wants to focus largely on global reforestation efforts for the purpose of fighting climate change. Their specially designed seed pods are made out of biochar, potentially giving them a leg up on overall CO_2 sequestration numbers. Company founders hope to plant 100 million trees per year by 2023.[13]

UK-based drone eco-restoration company Dendra has pledged to plant an incredible 500 billion trees by 2060. Dendra's drones are capable of planting 120 seedpods per *minute*, roughly 150 times faster than planting by hand. Dendra's plan is reportedly to employ 400 teams of two operators flying ten drones apiece to plant billions of trees annually (and at a much cheaper cost than planting manually).[14]

Or how about using drone tech to inspect forest topographies for signs of tree growth limitations? One research team has reported using drone equipment to survey encumbrances on tree growth, like woody climbing plants, which are essentially plants using trees as scaffolds to grow.[15] These plants can ultimately eclipse parts of a developing tree from sunlight, restricting its growth over time. This hindrance then reduces the amount of carbon forests can store, particularly in tropical areas. Using advanced technology to detect and potentially remove these tree growth restrictions will allow us to further increase the carbon storage potential of forests.

Forest health will take on even more mounting importance, given the stakes of keeping all these new trees alive. New scientific assessments of forest qualities for best practices can help. For example, "structurally complex" forests in which vegetation arrangements are highly variable seem to be better at carbon sequestration than others.[16] Researchers were able to use laser-based topography analysis tech like lidar to conclude that increasingly variable leaf arrangements and species was correlative with increased carbon storage in soils.

Meanwhile, cross-disciplinary research has helped researchers understand the role that safeguarding vertebrate species can have in fostering stronger forests—and hence greater carbon storage.[17] We now more fully understand the role that non-native invasive pests and diseases can have in weakening carbon storage potential. When researchers analyzed 430 non-native insects and diseases in US forests, they found the carbon transferred to dead materials (rather than into root systems) was comparable to the emissions from roughly 4.4 million cars.[18] Research also tells us that tree species more native to soil topographies will have a higher capacity to sequester carbon in general.[19] These, among other research findings, all help push the needle in keeping newly forested areas thriving.

Planting trees doesn't require a high-tech, guns-blazing approach to make a difference. But the option helps, even if campaigns to make it happen the old-fashioned way are ongoing worldwide.

Between the years of 1990 and 2015, for example, European countries were able to reforest an area the size of Portugal. Many more campaigns are now underway to improve upon that. England has recently declared it intends to plant 1 million trees by 2022. Ireland has pledged to plant 440 million trees by 2040, citing benchmarks of 22 million trees per year over 20 years.

In 2016, the Indian state of Uttar Pradesh broke the Guinness World Record for one-day tree plantings with an astounding 50 million trees. A year later in 2017, more than 1.5 million Indian volunteers planted 66 million trees over just 12 hours. Neighboring Pakistan's so-called Billion Tree Tsunami ran from 2014 until 2017, producing over a billion trees spanning 350,000 hectares for a price tag of $169 million. Ethiopia would later shatter that record by planting an estimated 350 million trees in one day during 2019, closing schools and government offices to accomplish the task.

Multinational organizations have also joined the cause: in April 2019, international Sikh leaders implored believers worldwide to plant one million trees as a "gift to the planet." The UN-endorsed Bonn Challenge is going for 150 million hectares by 2020; 350 million hectares globally by 2030. We have the ongoing 20x20 Initiative in Latin America (20 million hectares by 2020), the AFR 100 (100 million hectares by 2030), the New York Declaration on Forests (350 million hectares by 2030), and more. The UN's "Trillion Tree Campaign" has already begun with an estimated 15 billion trees planted around the world to date at the time of this writing.[20]

Australia has pledged to plant 1 billion trees by 2030. The island of Madagascar pledged to plant 60 million trees throughout the year 2020 alone. The Philippines has actually signed legislation mandating that students need to plant 10 trees apiece before they can legally graduate from a tier of education. That means 10 trees for elementary school, 10 for high school, and 10 for college. In all, officials estimated that about 175 million trees could be planted annually through the initiative. And citing a reasonable survival rate of about 10 percent, that's still an amazing 17.5 million viable trees each year.

Progress in growth can take many years to accumulate, but the results can be incredible. Over the past 30 years, Costa Rica has literally doubled its forest cover. China's anti-desertification program, also known as the "Great Green Wall," has planted more than 50 billion trees since the 1970s. In February 2019, the country even dispatched tens of thousands of its army personnel to plant trees in a bid to fight pollution. The decision augmented China's larger goal of planting about 84,000 square kilometers of trees—roughly the size of Ireland—by the end of 2019.[21] State media reported that 23 percent of the country's total landmass would be earmarked as forest cover (the goal is for 26 percent by 2035).

The debate about whether to re-seed forests specifically lost to wildfires still carries on. But at the very least, even if rapidly afforested areas do diminish due to wildfires, research indicates that a nontrivial amount of carbon is still sequestered by the trees that end up burning. The charcoal produced by fires (so-called pyrogenic carbon) helps to compensate for some of what is lost in emissions. After all, wildfires tend to burn an area roughly the size of India around the globe in an average year,[22] so we can expect some burning to occur.

Afforestation projects also become less meaningful if richer countries simply finance efforts in poorer countries while both nations count the carbon credit. These are the kinds of nuts and bolts issues we have seen countries trying to navigate since the Paris climate agreement was drafted in 2016 at the UN. Debates are again ongoing.

Countries with the most available landmass clearly have the most potential to add trees. Researchers with ETH Zurich conducting the aforementioned landmass availability study cited Russia at the top of that list, followed by the US (with 103 million available hectares), Canada (with 78.4 million hectares), Australia (58 million hectares), Brazil (49.7 million hectares), and China (40.2 million hectares).[23] The World Resources Institute has further estimated that an area roughly twice the size of Canada is available if we're counting forests being logged or degraded.[24]

Even with all this land, all these plans, and high-tech seeding, tree-planting campaigns need to begin immediately to be the most effective. Researchers realistically estimate that we won't fully reap our carbon's worth from afforestation efforts for roughly 50 to 100 years.[25] We will certainly benefit from foresting efforts far before then, but that is about the benchmark for maximum reward.

Many landmass estimates for available reforestation areas have focused on terrestrial regions, but that bypasses a crucial part of our afforestation portfolio. Just consider blue carbon. Mangrove areas, salt marches, and seagrass meadows have proven themselves superior "carbon sinks" and should be part of our silver buckshot. You will often find blue carbon tree systems lining the coasts of tropical areas, creating an extensive root system already visible above sediment in some areas. These areas are prized for carbon storage because the decaying organic matter containing absorbed CO_2 will sit buried under the watery sediment below. Once there, oxygen is less likely to release it through decomposition. Incredibly, Deakin University marine ecologist Peter Macreadie once estimated that blue carbon ecosystems can actually store carbon about 40 times *faster* than fully terrestrial forests.[26]

Mangroves do have a unique history of underestimation and overestimation for their sequestration abilities, but new research seems to indicate that they're nearly as good as blue carbon scientists thought. For example, the world's mangrove sites have been estimated by Doctors Robert Twilley and Andre Rovai of Louisiana State University to sequester about 24 million metric tonnes of carbon into soil per year.[27] By their estimates, the wood and soil of mangrove forests along the world's coastlines currently hold roughly 3 billion metric tonnes of carbon (more than tropical forests).[28] In Australia alone, the seagrass, mangrove, and salt marsh carbon storage has been estimated to trap about 20 million tonnes of CO_2 every year—roughly the emissions equivalent of 4 million cars—essentially locking it away in the soil there for millennia to come.[29]

Restoration efforts are immediately necessary, as the loss of mangrove forests alone has been estimated to produce 24 million tonnes of otherwise trapped CO_2 emissions each year according to one study published in *Nature*.[30] Simply stopping this loss would be even cheaper than active restoration efforts. Sadly, researchers have provided cost estimates comparing entirely new restoration efforts versus simply paying people who deforest to re-plant instead (with the latter proving to be more cost-effective, as painful as that sounds).[31]

The high-powered carbon sink abilities of coastal wetlands are great for several different reasons. Chief among them is that finding a significant source of carbon sink will offer us a powerful tool to help store atmospheric carbon. But it also means that organizations and companies working with blue carbon will have a "small investment, high reward" scenario for carbon storage. Companies and governments who wish to boast about how much carbon they've sequestered can pay far less than traditional tree-planting campaigns and achieve greater marketable gains. In April of 2019, for example, Apple announced a partnership with Conservation International to restore and then protect a 27,000-acre mangrove forest in Cispatá Bay, Colombia. They estimated that the forest will sequester roughly 1 million metric tonnes of CO_2 over the span of its lifecycle. Companies calculating as little payment as possible for max CO_2 storage might sound poor in principle, but the effect is still immense.

We do have other aquatic carbon capturing plants to discuss in our blue carbon portfolio, but their impacts can be less obvious. Seaweed, for example, is a macroalgae capable of capturing carbon just like terrestrial plants do. How much could "marine afforestation" help us in terms of sequestering carbon if deployed on a large scale? Researchers from the National Center for Ecological Analysis and Synthesis noted that available areas suitable for cultivating seaweed worldwide tally up to about 48 million square kilometers, and if even just a tiny portion of that area were farmed (approximately 0.001 percent of it), the results would be enough to render the world's aquaculture industry carbon neutral.[32] In fact, researchers estimated that cultivating seaweed in just 3.8 percent of the American West Coast Exclusive Economic Zone (a marine area extending no more than 200 miles from the coast in which a country has rights to natural resources) would be enough to offset the carbon produced by California's agriculture sector entirely.

Brian von Herzen, Executive Director of the Climate Foundation and giant of marine permaculture research, has developed fascinating new marine permaculture arrays (MPAs) to drive this effort forward. The arrays are essentially latticed structures submerged a few dozen feet below sea level for kelp to latch onto. Buoys attached to the structure rise and fall with passing waves, pumping colder and nutrient-rich waters up from below. The kelp consumes those nutrients, grows, and helps establish a network of additional marine animal life around it. Kelp can either be left to sequester CO_2 and then sink to the bottom of the ocean, or it can be harvested for biomass.

Researchers are still uncertain about just how much macroalgae like seaweed could contribute to global CO_2 removal and storage over a long timeline. For that reason, some scientists have explored seaweed cultures as more of a carbon-light nutrient source that doesn't need to compete with land-based options. In fact, one paper recently explored how we could use a local BECCS operation to power algae farms for an energy-positive, carbon-negative outcome acquiring plant food with no competition against land use.

One additional option for aquatic carbon storage through photosynthesis exists in phytoplankton, a microalgae also living on sunlight and consuming CO_2. Phytoplankton consume CO_2 on a level roughly equivalent to other algae species and land plants, but much of it is left in near-surface waters when the organisms decompose. Some of the consumed CO_2 falls to lower depths in the ocean, however, presenting us with an opportunity for a more literal carbon sink.

Ocean-Based Climate Solutions in Santa Fe has created a device that stirs up water in the ocean not only to promote the growth of phytoplankton—encouraging more carbon dioxide consumption—but also to "downwell" surface-level particulate matter deeper into the ocean. The result is a dual increase in phytoplankton gobbling up CO_2 and the recycling of more into the lower depths than would otherwise sink there. The team plans to make money by selling a "stock for carbon" design in which companies may sponsor the removal of CO_2 they emit through Ocean-Based technology. As of now, their carbon price will be $6.14 per ton (so comparatively very cheap, relative to some other removal options).[33] While new research tells us that previous models of phytoplankton carbon sequestration have been a little too optimistic,[34] the option still appears to be a viable sink.

Scientists tell us that bamboo is also worth considering. Bamboo is one of the fastest-growing plants on the planet, sequestering CO_2 faster than nearly any other plant.[35] Various civilizations have utilized bamboo materials as a substitute for timber in everything from furniture to clothing, while newer advancements have opened up more possibilities. Bamboo can thrive in a range of environments and even re-sprout after harvesting without the need for re-planting. According to Project Drawdown, the sequestration rate of bamboo is 15.4 tonnes of carbon per hectare per year (which is truly quite remarkable). They therefore estimate bamboo will reach a distribution rate of over 100 million acres by 2050.

Afforested acreage, tonnes of carbon sequestrated per hectare per year . . . realistically, many factors could change our afforestation estimates. The species of tree or crop being planted matters. Whether or not tree plantings coincide with carbon crops matters. The amount of available land could also shrink over time as a direct result of soils impacted by climate change holding less moisture. Some areas identified as afforestation candidates could even be underwater. By one study's estimate, the global "potential canopy" may shrink by approximately 223 million hectares by 2050. We're talking about real land lost due to climate pressures.[36] Alterations in Earth's hydrological cycle due to climate change will also make wet years wetter and dry years dryer, wreaking growth havoc on particularly sensitive tree species. Research tells us that increases in rainfall extremes can potentially lead to long-term declines in tree growth.[37] Continually analyzing these factors will matter as we scale up our afforestation efforts.

Now, this is where afforestation gets even more interesting. Because while we may lose some land where we'd hoped to place new trees, advanced new sciences can help us do more with the space we have. Consider the new developments scientists have achieved with bioengineering new biofuels and industrial materials. Can we possibly help improve our outlook on afforestation efforts through biotech? Can we potentially utilize genomics and synbio goals, like researching the particular genes in plants that code for superior CO_2-fixing or root-storing mechanisms—isolate those genes—and then genetically modify new crops, plants, or trees to contain those advantageous carbon-capturing genetics?

Enter Dr. Joanne Chory, a biologist studying plant genetics at the Salk Institute in San Diego. Along with her Harnessing Plants Initiative team at Salk, Chory and her lab are hard at work biohacking plant lifecycles to favor increased CO_2 storage within larger root networks.

The Chory lab's plan is to engineer common crops to establish bigger, more diffuse roots full of suberin, a waxy material that is part of corks. Because suberin can store carbon within a plant's root system and is much more resistant to decomposition, the result could be plant life effectively "supercharged" to store carbon. Same amounts of land; more carbon potentially stored.

A deeper root system will ideally allow CO_2-packed roots to remain in the soil even after farmers harvest crops at surface level, sequestering the stored carbon for potentially hundreds of years. The overall result would be an increased amount of CO_2 storage through crops grown to feed world populations anyway, bringing a powerful new option to our carbon-management portfolio. Chory and her team estimate that deploying crops like these into the global agricultural food chain could reasonably lead to about a 20 to 46 percent reduction in CO_2 emissions each year.[38]

I reached out to the Chory lab while compiling new research for this book and the exchange was encouraging. As of this writing, the team is nearly one year into a five-year research effort. They will conduct two years of lab research and two years of greenhouse trials before moving into real-world fields for their final year of testing. Progress has reportedly been excellent thus far in the lab phase, so we may have great results to look forward to.

On paper, the team knows what they're looking for: genetic pathways for increasing suberin, enlarging root systems, and creating deeper-growing root systems. And if trial phases are successful, the team may be able to utilize their techniques for staple crops like corn, soybean, rice, wheat, cotton/cottonseed, and rapeseed/canola. Better yet, deeper and stronger root systems could also help protect land from erosion.[39]

Nobody's quite sure yet whether suberin or root augmentations will create a longer developmental arc for plants to reach maturity. Or if the crops will taste any different than they normally would. But while the team has been easing into this research with traditional plant-breeding techniques, CRISPR applications are not entirely out of the question. So even as this research is just now beginning to ramp up in 2020, largescale commercial production could be rolling out in less than five years.

Are there other mechanisms within a plant's lifecycle we can potentially augment to increase carbon capture and storage?

Earlier, we discussed the potential for engineering photosynthesis itself. Alterations like those could help to impact carbon "fixing" efficiency.

With that in mind, professor Tobias Erb and his colleagues working at the Max Planck Institute for Terrestrial Microbiology in Marburg recently developed a synthetic metabolic pathway for plants. The research was undertaken to experiment with the way plants encounter and process

CO_2 generally, with an aim toward both speeding up the process and making it less energy-intensive.

Normally, plants utilize an enzyme called Rubisco (for short) to capture carbon dioxide, but the work is slow and mistakes can be made. Rubisco reportedly captures an oxygen molecule instead of CO_2 in one of every five reactions.[40] Erb and colleagues intended to go searching for a more successful version of this enzyme and potentially develop a genetic scaffold to work it into test plants. The team spent more than two years analyzing every individual biological component of the metabolic cycle, including all the enzymes involved—and then sifted through about 50 million genes and 40,000 enzymes to settle on a winning formula. The result? A synthetic metabolic pathway utilizing a turbo-charged version of Rubisco found in a strain of bacteria that does not make the same mistakes throughout CO_2 fixation.

If successfully integrated, the team's synthetic metabolic pathway will convert CO_2 absorbed from the surrounding air into organic matter more efficiently than plants manage via traditional photosynthesis. According to researchers, their synthetic metabolic pathway could potentially fix carbon dioxide from the atmosphere 20 percent more efficiently than plants can normally.[41] This may alter our afforestation prospects for negative emission efforts. Even newer research from a bioinformatics team at the University of Würzburg has reinforced the prospect that a modulated metabolic pathway could enable plants to absorb five times more carbon dioxide than in their normal state.[42]

Or how about speeding up the CO_2 absorption process? Scientists at the University of Glasgow recently developed a new, synthetic method of speeding up the opening and closing of plant stomata—pores in the plant's leaves through which CO_2 is drawn. To do so, scientists utilized a technique called optogenetics—an incredible new means of directing cellular behavior through the effect of lights on ion channels. By inserting light-activated genetic coding into an organism, scientists can impact cellular metabolism in ways that speed it up or slow it down through variations in light exposure. Essentially, within the context of altering plant cells, synchronizing stomata activity with light through a genetic switch has enabled scientists to dictate not only increased plant growth, but water conservation and—more importantly—increased CO_2-fixing.[43]

Scientists weren't necessarily picturing climate change applications in this research, but the options for compounding the effects of photosynthetic approaches in mitigating climate change may, at some point, be helpful. And as a side note, because this synbio research focuses on light and plant biomass growth, developments like these could really be game-changing for indoor vertical farmers as well. While we cannot often manipulate the light levels present outdoors to alter stomatal kinetics, we can indoors. Scientists may therefore be able to develop better light recipes reducing water needs while maximizing biomass accumulation and overall growth.

ARTIFICIAL PHOTOSYNTHESIS:
MECHANICAL AND BIOHYBRID VARIATIONS

Buckle your seatbelt.

The complicated truth about traditional afforestation is that researchers still speculate whether these efforts will be enough to accomplish what we want. For a sense of scale, ~10 Gt of carbon dioxide removed per year was the benchmark for drawdown efforts from the year 2050 onward. Drawing down just 1 Gt of carbon dioxide through traditional agriculture would require roughly a million square miles to accomplish.[44] That would equal roughly the combined surface area of Germany and France. We therefore will require incredible amounts of land to accomplish our carbon goals through traditional photosynthesis and plant life.

Fortunately, conventional plant life is not the only photosynthetic game in town. The carbon-absorbing capabilities of algae and cyanobacteria can also fit into these discussions, along with the artificial photosynthesis tech discussed throughout this book. And what makes these respective biohybrid and mechanical approaches so attractive are the commercial prospects driving these techs forward.

Consider the "biosolar leaf," for example. Biosolar leaf technology takes advantage of our potential ability to arrange colonies of other photosynthetic organisms into compressed configurations the same way one might plant a forest. And we may be able to achieve a *greater* amount of net carbon storage than traditional plant life can manage in a similar space.

Until now, tubes teeming with tiny little photosynthetic organisms have been discussed relative to the chemicals the organisms are producing therein. Algae produces biofuel, bioplastics, and more. But we can also do carbon management through these technologies. If we consider comparing a forest of trees versus a forest's worth of encased microalgae, we can then assess whether one option may make more sense in some situations for space used, resources required, etcetera. Particularly as capsules of microscopic algae can essentially be placed anywhere on Earth, including places where trees do not often thrive.

Researchers at the Imperial College in London working in partnership with new British start-up Arborea revealed the world's first-of-its-kind biosolar leaf-shaped technology back in early 2019. The equipment they envisioned was basically a solar panel-like structure covered with encased microalgae and phytoplankton. These bright green panels could be placed on rooftops, the sides of buildings, or even in the desert. The prototype model Arborea used to demonstrate this technology was eventually shaped like a chandelier—with a beautiful, green, chic design for indoor placement and the added benefit of providing more oxygen to the surrounding air. Arborea estimates that their biosolar leaf panel can potentially accomplish the work of 100 trees while only using the physical surface area of one.[45]

Tubes packed with microscopic algae will do more than allow us to draw in CO_2 and clean the air basically anywhere we have access to sunlight. We can also harvest the biomass produced by algae thereafter, yielding an end product of edible plant protein. And therein lies the entrepreneurial incentive for rolling out a product like the biosolar leaf. As the encased microalgae nears the end of its lifecycle, the tubes can be drained. The resulting material can be processed into what will eventually become a nutritious, meal-supplement powder. If you picture what Western health marketers have done with a product like spirulina drink mix, you're essentially there.

While algae (macroalgae like seaweed in particular) has been a staple of many Asian countries for centuries, Western countries looking for more sustainable sources of antioxidants and protein may finally be ready for it. Technology like this may one day be placed, as Arborea envisions, along the sides of large buildings and on rooftops—bright green casings filled with microscopic life lapping up CO_2 and producing food in the process. And better yet: using pre-captured CO_2 and upping the concentrations for enclosed algae could even accelerate their growth rate.

Arborea is not alone in developing biosolar leaf technology and equipment like it. Mexican start-up BiomiTech has developed a structure resembling a lamp post (again, able to be placed anywhere) in which encased microalgae sits where the light portion would normally. The BioUrban, as they have called it, stands roughly 14 feet tall and weighs about a ton. It is reportedly capable of scrubbing as much pollution from the air as 368 trees. You'll be paying for it, though: each BioUrban unit costs roughly $50,000 American dollars at the time of this writing.

Austin-based Hypergiant has created a similar piece of equipment with its Eos Bioreactor. Their 3x3 standalone bioreactor houses algae in compartments utilizing AI to direct a more focused growth cycle. Machine-learning algorithms tailor a series of health monitoring systems to adjust the amount and type of light reaching algae; the amount of CO_2 entering; temperature; pH; harvest cycles; and more. Harvesting the biomass created by algae inside the reactor is rather like Arborea's work, although Hypergiant offers some of the other refinements taken from the biofuel and synbio sections of this book—with biomass potentially ready for fuels, oils, fertilizers, plastics, and other goods. Moreover, Hypergiant has committed to keeping their blueprints public in the hopes that online DIY creators can improve upon community and residential-level units. In all, just one of Hypergiant's bioreactors is reportedly about 400 times more effective at absorbing carbon than trees using the same space. An image of Hypergiant's bioreactor prototype design can be found on page 162 in the photo section prior to Chapter 8.[46]

What helps to make these prospects so alluring is that bioengineers may one day augment algal strains in ways that make some or all of these ventures more viable.

Amsterdam-based Photanol and Vancouver-based Phytonix, for example, both harness the power of engineered cyanobacteria to thrive on sunlight and absorb CO_2 to create valuable products in the process. In some cases, the cyanobacteria will create bioplastics. In others, it's butanol through the same process mentioned earlier in the EU's photofuel research campaign. In the case of Phytonix, fuel is again excreted into the medium in which cyanobacteria live. Along with the cheaper butanol the company is able to sell, Phytonix claims the whole process is carbon negative.[47]

A hybrid photosynthetic technology like the biosolar leaf and others like it represent a great option in our climate response portfolio. But can we do even more through photosynthesis—capturing carbon and harnessing sunlight to produce energy—entirely without photosynthetic life?

To recap, the artificial photosynthesis process takes the energy in sunlight and splits water (H_2O) into hydrogen (H_2) and oxygen (O) while using carbon dioxide to help augment this process. The technical term for artificially "fixing" CO_2 the way plants do is called light-driven carbon dioxide reduction. Essentially, the sun's energy causes photoexcitation of another molecule (a photosensitizer) that can absorb light. This interaction creates an electron shift within the absorbing material that draws a subsequent reaction (read: reduction) from surrounding CO_2. Ultimately, continuing in the process of artificial photosynthesis, these molecules are then shuttled to either isolate the hydrogen obtained or feed it to engineered bacteria. We discussed this earlier in terms of what the end-goal products were: energy or biomaterials for refinement. But just how effective could this process be in converting CO_2 on an atmospheric scale?

Sadly, we are not yet technologically sophisticated enough to use artificial photosynthesis for long-term atmospheric capture. By some estimates, however, while the bacteria involved are producing bioplastics like PHB, they are reportedly able to remove about 180 grams of CO_2 from the air for every kilowatt-hour of solar-generated electricity used.[48] That is admittedly not a whole lot of CO_2, but it is bolstered by the fact that they're doing it essentially for free and on renewable energy.

In terms of technical efficiency, artificial photosynthesis can reportedly outperform normal photosynthesis by a wide margin. Some scientists, like Matthias May, a physicist at the Helmholtz-Zentrum Berlin Institute for Solar Fuels, have run rudimentary calculations on the effectiveness of using artificial photosynthesis for carbon removal. May estimates that the effectiveness of this equipment in terms of landmass—for removing our required 10 Gt/yCO_2 by mid-century—would be roughly in the area of 27000km². That would be a considerable improvement over the million square miles cited for traditional agriculture (even if it would be a truly incredible engineering feat to produce). An area the size of Hawaii could theoretically capture as much carbon dioxide as an area the size of Europe "covered in the most carbon-hungry plants," according to May.[49] And unlike a forested area, solar-powered artificial photosynthesis could also produce fuels and industrial materials in the process.

I corresponded with May while researching what widespread artificial photosynthesis ops might one day look like. He advised me that we still have plenty of work to do regarding the development of stable catalysts that we can use at the abundance and scale of atmospheric carbon capture. May also explained that engineers are still working out the photo-electrochemical interface in how the equipment absorbs and outputs the CO_2 we seek. He believes a largescale facility would also make less sense than spreading tech like this out over various distances (so, possibly the space of Hawaii spread out around the globe, for example).

Chief among our problems in using artificial photosynthesis for capturing atmospheric carbon is that much of the equipment has typically been used indoors. One exciting new development from researchers at the University of Illinois at Chicago has focused on placing the equipment within a semi-permeable membrane to allow water inside—which then draws in CO_2 as it evaporates and moves through the membrane. This development is interesting for two reasons. For one, it more closely resembles an actual leaf in continuously absorbing CO_2 along a moisture gradient, rather than the more dichotomous on/off operation of other capture equipment. But more importantly, it has allowed artificial photosynthesis to be more outdoor-focused (where this tech realistically needs to be). With it, we move closer to a truer-to-form artificial leaf.

By the authors' calculations, a 500m² area of their artificial leaves—each 1.7 meters long by 0.2 meters wide individually—would be able to reduce carbon dioxide levels by about 10 percent in the surrounding air within 100 meters of the array in one day.[50] The technology is modular (able to be built-upon incrementally), scalable, and is approximately 10 times more efficient than natural leaves at converting carbon dioxide.[51]

While it's hard to match the price that trees charge for performing photosynthesis (it's free!), artificial leaf tech may help us in the instances where land considerations or time scales require that we explore it. Research in the area is continually evolving. I was recently able to speak to another artificial leaf engineer, professor at the University of Waterloo (Ontario) Yimin Wu, regarding his own work in this area. Wu's research capitalizes on avoiding the pitfalls of other tech variants using multistep processes and expensive, relatively rare catalysts. Wu's process, on the other hand, involves just one step and uses a much cheaper, Earth-abundant material: cuprous oxide. When I spoke to Wu about his research and asked him whether drawing atmospheric CO_2 into fuel production at vast scales will ever be achieved in the near future, he quickly replied: yes.

SCRUBBING CARBON FROM THE AIR:
DIRECT AIR CAPTURE

Let's recap our carbon management options thus far. We have the power of photosynthesis with plant life. We have the extended use of photosynthesis in biohybrid techniques with microalgae and cyanobacteria. We also have artificial photosynthesis techniques driven by sunlight that involve using non-photosynthetic organisms. In those cases, solar-powered electrolysis shuttles hydrogen to an on-board microbe for consumption and metabolism. We additionally have artificial photosynthesis techniques operating entirely without organisms at all, but still utilizing a catalyst driven by sunlight to convert/reduce CO_2. So what, exactly, is direct air capture?

Unlike artificial photosynthesis techniques, direct air capture equipment operates through a combination of fans, filters, and solvent and sorbent materials.

The primary benefits of using DAC technology to capture carbon are that the space used and speeds of capture are far superior to the photosynthesis approach. The reality of atmospheric-level capture currently falls short of the promise, however. Many direct air capture facilities are still relatively reliant on fossil fuels. Moreover, the energy spent capturing carbon is comparatively high, meaning that facilities must off-set their own emissions even as they price captured carbon at a competitive rate. Building and maintaining facilities also can carry a number of environmental costs. But perhaps the most debilitating factor is that the market for captured carbon is still not very well developed. So companies either need to refine their captured CO_2 into something useful or sell off the "carbon credit" of captured CO_2 to an industrial emitter.

In current typical capture machines, large fans pull wind through a chamber to separate the CO_2 inside. Once inside, the carbon dioxide in ambient air moves across a series of strips in a sorbent or solvent material. These strips contain negatively charged ions which bind with the gas molecules, converting them into carbonate material. Capture machines then saturate those strips with water inside a separate compartment, "scrubbing" the CO_2. Draining that water then creates an environment in which the carbon dioxide can reform, suddenly more concentrated thereafter. Engineers can then suction that CO_2 out into an adjacent tank.

Newer capture technologies have since been invented, but most of the major capture companies now in operation use sorbent materials. For reference, there are about 40 largescale carbon capture installations throughout the Americas, Europe, Asia, and MENA at the time of this writing. We also have so-called "point-source" capture with equipment designed to cut emissions at the source.

The US currently has the most carbon capture operations fully operational. American carbon capture plants also currently outperform the rest of the world by a country mile: the Century Plant in Texas captures about 8.4 $MtCO_2$ per year, while the Shute Creek Gas Processing Facility in Wyoming captures about 7 $MtCO_2$ per year.[52]

Norway was the frontrunner of largescale carbon capture tech with its offshore Sleipner gas field in the North Sea (going as far back as 1996). The facility captures CO_2 from ambient air and injects it into sandstone formations under the sea, ultimately capturing about 1 million tonnes of CO_2 per year. Norway has also rolled out several installations since, capturing about 400,000 tonnes of CO_2 annually apiece.

South Korea has two exciting new facilities in development, which could be operational by the time of this book's publication. The Boryeong KoSol CO_2 Capture Test Site and Hadong Dry Sorbent Capture Test Site are both testing out the possibility of using thermal power to drive capture operations. Both are advanced versions of typical capture sites operating elsewhere, the former using amines and the latter using dry sorbent.

Carbon Engineering in British Columbia is a particularly compelling capture company. Run by Harvard physicist David Keith, Carbon Engineering has led the industry to a milestone low in costs for operation. While industry critics have routinely assessed the capture industry as economically unfeasible, Carbon Engineering has meticulously reduced potential costs to a marketable level. As other companies around the world were capturing carbon for about $400–600 per ton, Keith and his team were able to bring this amount down to the magic number of $100 per ton; a number market analysts had previously stated would be necessary for competitive carbon capture.[53] They're still conducting trial runs as of this writing, but the prospect is there. You will find an artist's rendering of how Carbon Engineering may one day deploy direct air capture fans within the latter photo section on page 163.

The fledgling industry of carbon capture will benefit immensely from the American tax provision 45Q incentivizing capture, which was signed into law by the Trump administration. I suspect the administration (or industry lobbyists, perhaps) had seen the writing on the walls in terms of putting a price on carbon. They won't come right out and say it, but providing cash for industries that will capture their carbon is essentially the same as taxing those who don't. The provision gives $50 per metric ton of CO_2 captured and stored, meaning that companies have an economic incentive to lower emissions.

Carbon Engineering's first commercial plants will reportedly capture one million tonnes of CO_2 per year apiece.[54] That means one facility will be able to perform the carbon capture work of about 40 million trees, sequestering the annual emissions equivalent of roughly 250,000 cars.[55] Sounds worth pursuing, right? And because CO_2 is distributed fairly evenly around the atmosphere, facilities can essentially be placed anywhere.

As a potential solution for drawing down CO_2 and reversing climate change, direct capture facilities certainly offer us the space and the power. If we were to power direct capture facilities entirely with renewable energy sources, we could slowly capture carbon to either sequester underground or use in synthetic fuels and manufacturing materials. But unless there's a business model backing up this process, the public will be footing the bill—potentially a very high bill with the amount of direct capture plants that would be necessary. Moreover, any DAC venture will be essentially useless if heavy emitters are still pumping CO_2 into the atmosphere.

Swiss company Climeworks is another largescale capture company, with 14 facilities operating throughout Europe and more on the way. Climeworks capture machines look somewhat like jet engines, drawing in air to force it through capture equipment. Company representatives told me that each "collector" is able to capture about 50 tonnes of CO_2 per year. Climeworks typically places 6 collectors within an overall container, which is often what businesses install. The business is thinking big, though: their latest and largest installation, slated for late 2020, will feature 30 to 40 containers in a Reykjavík facility.

It's easy to see the appeal of companies doing business with Climeworks. With expansions on the way, Climeworks has plans "to offer individuals, countries, businesses and institutions the opportunity to reverse their past, present and future emissions, permanently and safely."[56] Just one rooftop unit at the company's Hinwil location in Zurich is reportedly able to remove the carbon equivalent of about 36,000 trees.[57]

Professor Klaus Lackner's "mechanical tree" equipment is also capable of capturing CO_2 at atmospheric levels. Unlike other direct capture machines, which physically draw air through the equipment, the artificial tree simply lets wind blow CO_2 through the system passively. The equipment itself contains stacks of sorbent-filled disks. Air flow moves CO_2 across the discs where the capture occurs. Discs are then lowered to a separation chamber where the CO_2 is released from the sorbent. Each "tree" is capable of removing 1 tonne of CO_2 per day, with full-scale farms of trees aiming to capture 100 metric tonnes per day.[58] Lackner's technology is reportedly thousands of times more efficient at removing CO_2 from the air than normal trees (not in gross amount of CO_2, but they still surpass normal trees there as well).[59]

As the Director at Arizona State University's Center for Negative Carbon Emissions, Lackner imagines that we may one day have vast forests of these mechanical trees lining the countryside, soaking up billions of tonnes of carbon dioxide.[60] Not a pretty sight, if you're a superfan of natural trees. Particularly if the equipment generates a constant hum due to the recycling of carbon-soaking surfaces required for scrubbing. And on a pragmatic level, we're talking about building the requisite infrastructure: foundations, maintenance roads, and disrupted natural systems.

There lies the rebuttal from folks who do not see a point in DAC technology even if it is renewably powered. Critics argue that even if DAC facilities can be powered without emissions, there will still be a significant toll taken on the environment for construction and maintenance thereof. It is tough to validate that argument, however, if DAC tech were simply one more pellet of silver buckshot operating with zero-emissions energy. And DAC technology represents a realistic prospect for helping to decarbonize otherwise hard-to-decarbonize industries.

A second argument against DAC is that it also potentially diverts money away from developing renewables, which would be more CO_2 we wouldn't have to scrub out of the atmosphere—a 100 percent capture rate, in other words.[61] However, I believe this ignores the pre-existing, incredible amounts of CO_2 in the atmosphere that even a zero-emissions and carbon-neutral world would struggle to draw down in time.

Internationally renowned MIT engineer and all-around mental giant Howard Herzog has been a vocal critic of direct capture technology. I corresponded with him on the current state of this technology, and his response was telling. He believes energy spent on capture equipment is still currently debilitating in a business model, and is unlikely as a primary means of drawing down atmospheric CO_2 to save ourselves.

That last point is particularly crucial in DAC discussions. While we suffer from an excess of CO_2, we still have a very large atmosphere and the parts of CO_2 within it are still fairly small. There is a difference between scrubbing CO_2 from smokestacks versus the Earth's atmosphere. To actually put a dent our atmospheric CO_2, we would need an incredible number of DAC machines operating at today's standards. Glen Peters, a research director at the Cicero Center for International Climate Research in Oslo, estimated that we would need to build a large direct air capture facility like those of Carbon Engineering or Climeworks thousands of times over.[62]

Worse yet, if today's direct air capture methods are to be used at a meaningful scale for atmospheric carbon, this also means incredible amounts of sorbent and solvent materials for the equipment to use. This will potentially create immense amounts of waste disposal.

So where's the good news?

The good news is that like the biohybrid photosynthesis approach, market forces may soon propel this industry forward. New developments are continually making capture equipment more efficient. Admittedly, though, some of them get a little complicated.

Companies like Carbon Engineering capture CO_2 by forcing air through an alkaline liquid solution. The CO_2 dissolves into the liquid and forms carbonate. CO_2 is then extracted back into a concentrated gaseous form . . . and then ultimately into the chemical building blocks of fuels and plastics in equipped refineries. One method of carrying out this process is to convert carbonate into a solid salt and then heat it to temperatures topping 900°C. The materials then undergo further transformations.

The energy required to carry out typical direct air capture can be very expensive. But researchers at the University of Toronto have found that by adding a uniquely formed electrolysis design to the component involving a dissolved liquid solution, they could convert carbonate directly into syngas—skipping the heating step and saving significant amounts of energy.[63] The innovation, then, refines carbonate to syngas in a single step—a development improving the economics of capturing carbon by eliminating some of the energy-intensive losses. And because the technology and research are still so underfunded, new developments inching the tech toward widespread viability could crop up at any time.

Another new development in bypassing capture material expense focuses on alternative methods of filtering CO_2 molecules from ambient air. Normally, in other DAC designs, gaseous CO_2 must pass through a sorbent or solvent material first. But new research from scientists at the Chalmers University of Technology and Stockholm University have demonstrated impressive carbon capture with another material: zeolites.[64]

Zeolites are microporous aluminosilicate materials, typically used as industrial adsorbents. Zeolite particles have long been understood to capture carbon, but have historically been difficult to work with and implement in carbon-capture designs. The latest innovation was to suspend zeolites in a bio-based hybrid foam that acts as a lightweight filter. Small particles of zeolite are essentially suspended all throughout porous foam, capturing CO_2 as air passes through it with a high capture rate. As a highly porous cellulose and gelatin-based material, the foam is ultra-lightweight, sustainable, and durable with a high reusability rate to match its capture rate.

Incredibly, zeolites have also shown promise in direct air capture equipment for turning methane into carbon dioxide. It sounds counterintuitive because this entire chapter has been about drawing down and capturing carbon dioxide—but as a greenhouse gas, methane is much more powerful. Turning it into carbon dioxide could actually leave us better off, in terms of remedying climate change.

One daring new study has calculated that if we were to use zeolites to remove methane from the atmosphere, we could restore the atmosphere by removing "only" three billion tonnes of methane.[65] By these estimates, doing so would generate roughly a few months' worth of industrial CO_2 emissions. It would also eliminate up to a sixth of our current warming levels. Their technology is still a long way off, but it represents a curious new development in zeolite capture research.

New potential avenues for capture research are increasing the more these areas receive crucial funding. One recent development from other researchers at MIT has offered promise for directly capturing CO_2 from ambient air. Postdoc researcher Sahag Voskian and colleagues developed a new technique of passing air through a stack of charged electrochemical plates. The team's device is basically a specialized battery apparatus which absorbs CO_2 from passing air, whether ambient or concentrated. As air passes over the device's electrodes during its charging period, the gas is released during its discharge period. On the surface of each stack of electrodes, a compound with composite carbon nanotubes creates an electrochemical reaction with passing CO_2 during the device's charge— and then releases that CO_2 during discharge. That's essentially it: by alternating between charging and discharging, CO_2-affinity or not, air passing by is then concentrated into a pure CO_2 output (using a process known as electro-swing adsorption).[66] Voskian has produced a helpful animation online to illustrate how this all works.[67]

Crucially, Voskian's tech can operate at normal air pressures and at room temperature while many capture techniques require higher concentrations of CO_2. And with its all-or-none affinity levels based on charging or discharging, this device can do without some of the "intermediate" chemical processing steps and the heat requirements necessary in previous capture methods. The team hopes to commercialize this process eventually, and created a company called Verdox to do so.

Building a business model around selling captured CO_2 (or its refined product) will be of immense value for remedying climate change. Because if captured carbon has value, it can off-set the cost of installations for emitters and allow for a revenue stream that not only pays for installation and use, but also helps emitters avoid taxation. Let's move now to discuss some of the most exciting new developments in carbon capture: efforts to commercialize it.

We may still see the day in which subsidizing the capture of atmospheric carbon takes off. After all, the scale of simply funding carbon capture may eventually seem like a bargain compared to the disaster relief efforts of regions beset by superstorms, droughts, and floods imposed by an increasingly erratic global climate. David Hawkins, director of climate policy for environmental group the Natural Resources Defence Council, has compared carbon capture techs to an airbag for our species.[68] Debates about whether capturing our carbon may pose a moral hazard will likely drift to the background as more and more communities sustain damages from climate change. At that point, as Julio Friedmann, senior research scholar at the Center for Global Energy Policy at Columbia University has phrased it, carbon capture may then go from moral hazard to moral imperative.

Carbon capture need not be a quick fix, nor a means of perpetuating business as usual. Many climate reporters and environmental activists view direct capture as a "fig leaf" for dirty energy companies trying to convince us that the carbon status quo can continue. Threaded through the reverie for carbon capture is a fantasy of industrial absolution, writes the exceptionally insightful David Wallace-Wells in his work *The Uninhabitable Earth*— the fantasy "that a technology could be almost dreamed into being that could purify the ecological legacy of modernity, even perhaps eliminate its footprint entirely." Those of us who view carbon capture as a crucial tool in our climate response portfolio would do well to keep the skepticism in mind.

COMMERCIALIZING CO$_2$: THE CARBON ECONOMY

New carbon management technology will go nowhere unless we address a crucial question posed by capture critics everywhere: *who's going to pay for all this?*

I often want to reply that desirable public services are not usually free. That, in the words of Sir David Attenborough speaking at Davos in 2019:

> *We did not get to this point deliberately and it has happened astonishingly quickly . . . If people can truly understand what is at stake, I believe they will give permission to business and governments to get on with the practical solutions.*

Realistically speaking, however, the economy also helps to drive change. So if there's a market for captured carbon, we will have more support to "solve" the climate crisis. Good news, then: economists believe a marketplace for captured carbon dioxide is entirely possible.

Companies specializing in carbon capture currently survive by a mix of piping CO$_2$ to soft drink companies for use in fizzy drinks, off to greenhouses for crop growth, supplying pressurized gas for enhanced oil recovery by fossil fuel companies (oh, the irony), or by selling carbon off-set credits to heavy emitters. And while that might have been a long sentence, the market for captured carbon is still not very large.

The most comprehensive study published to date investigating the potential for various uses of carbon dioxide concluded that it could become a vast new global industry.[69] The researchers believed, in other words, that carbon capture and storage (CCS) will soon become carbon capture and utilization (CCU). Because if there's ever a practical enough market for captured carbon, why would we bury it underground?

Advanced new research has focused on the "upcycling" of CO_2 into other materials. And incredibly, up to 16 different carbon-based compounds can be refined from carbon dioxide, with potential for more products "downstream" from those. Various companies now specialize in producing these materials, or in something of a full-suite approach in offering heavy emitters the opportunity to generate a revenue stream from their waste CO_2. Berkeley-based company Opus 12 has produced a device to do precisely this.

The equipment Opus uses can essentially bolt to pre-existing sources of CO_2 emissions. Roughly the size of a washing machine in smaller operations or a modular shipping container in industrial settings, the Opus equipment allows emitters to recycle their CO_2 into cost-competitive chemicals post-emission. By using a proprietary catalyst, they can speed up the recycling process in ways many other capture techs cannot. The catalyst breaks apart CO_2 and water molecules into smaller components in the presence of electricity, which the tech can then re-form into new molecules. After the piped-in CO_2 is transformed to a new chemical product, the carbon monoxide (CO) can be captured and sold, or hydrogen can be added for further refinement. Simply involve the Opus tech somewhere along the route where CO_2 is expelled as waste and—using only water and electricity—transform it into market-ready chemical products.

In developing an economic scaffold for chemical compounds to be produced with recycled CO_2, otherwise emissions-heavy manufacturing will be reduced. Everything refined from recycled CO_2 is something that doesn't require as large a carbon footprint. Ethylene, for example, can be produced by the Opus tech. It is a precursor for most of the world's plastics. If carbon-heavy manufacturing is replaced, that's roughly 2 tonnes of emissions bypassed per year per ton of ethylene.

As a side note, companies like Opus 12 are already dreaming bigger with carbon capture technology. Opus cites the immense potential in using likeminded technology whenever we manage to reach Mars. Since the Martian atmosphere is 95 percent CO_2 and turning CO_2 into carbon monoxide (CO) consumes no net water, we already have a cost-effective way to produce oxygen on the Red Planet. The ethylene, meanwhile, could be used for 3D-printable tools and shelters. Power that with renewable solar energy and we're much more self-sufficient when we're up and running in space.

Crucially, though, while we may not always speak of a revolutionary market specifically *for* CO_2 which will galvanize capturing it for everyone, we can certainly speak of the businesses primed to potentially be carbon negative and support them. Companies like Arborea capture CO_2 while cleaning the air and producing edible plant proteins in the process. Seaweed farmers cultivate edible plant life from the world's oceans while capturing CO_2 in the process. Companies like Solar Foods produce sustainable protein in edible powder form with the added bonus of using captured carbon during production.

Consider a product from London, UK-based Photo.Synth.Etica: algal bio-curtains able to drape the sides of buildings with bright green, living algae. Bio-curtains debuted at The Printworks in Dublin Castle partway through 2018, looking like contemporary see-through green panels lit by fluorescent veins. Except—the curtains were alive. Each curtain was a photobioreactor encased in a bioplastic vessel teeming with microalgal cultures. Every $2m^2$ could reportedly absorb the same amount of CO_2 as a mature tree while doubling as futuristic community art.

Or how about a carbon negative vodka company? Brooklyn-based Air Co. begins by splitting water through renewable energy and obtaining hydrogen in the process. The hydrogen and CO_2 molecules are then mixed together using a proprietary catalyst (and presumably engineered microbes, I'm guessing) to form ethanol. That alcohol is then further refined with renewable energy in the company's Brooklyn distillery site into higher yields of ethanol. To ensure the product remains carbon neutral, though, you will have to be in New York City to procure it. In the meantime, you can see an ongoing tally of CO_2 captured on the company website.[70]

Several new lower-carbon industrial products were mentioned in the chapter on industrial replacements. Some of them could also apply here in a section about markets capitalizing on recycled CO_2. Consider Brooklyn-based 10XBeta, which boasts its "shoe without a footprint" with sneakers created from recycled CO_2 turned into Polyol chemicals.[71] Calgary-based Carbon Upcycling Technologies makes watches infused with nanomaterials derived from captured CO_2.

Or consider the work of Huntington Beach, California-based start-up Newlight, which captures methane or CO_2 from farm or power plant flue streams, then combines them with microorganisms designed to pull out the carbon components. That concentrated carbon is then combined with H_2 and oxygen to create a biopolymer plastic material. The company has since partnered with IKEA to produce sustainable bioplastics for the company's furniture lineup.

Consider also the incredible work from Chicago-based LanzaTech, a company that transforms waste gas emissions into ethanol fuel or consumer products with the help of an organism derived from synthetic biology. LanzaTech's engineered *Clostridium autoethanogenum* performs gas fermentation from the waste streams of industrial emitters, producing ethanol in the process or tailor-made upcycled chemicals. New developments in microfluidics and acoustic liquid handling have allowed LanzaTech's scientists to increasingly fine-tune the chemicals they can produce from waste gaseous streams, ultimately raising the possibility that more consumer goods (including materials like nylon and rubber) can be manufactured from recycled carbon. The company invites consumers to imagine a day in which planes are powered by recycled greenhouse gas emissions—and where yoga pants have started life as pollution from a steel mill, later upcycled into fresh material. Their "pollution to product" supply chain model further helps companies make more carbon-conscious decisions.

There's also a Canadian start-up called CleanO2, which uses a much smaller and residential-sized carbon capture device around the size of an air conditioner. Their device, CARBiNX, can be fitted to natural gas-powered boilers to reduce both energy use in commercial and residential buildings and lower greenhouse gas emissions in the process (up to around 10–20 percent).[72] CleanO2 also produces hand soaps and detergents with CO_2 mixed in. Consumers can choose products like these to help contribute to creating a carbon economy.

The concept of "locking away" CO_2 inside commercial products will be immensely beneficial for us going forward—even better if the integration of CO_2 makes a product even better than before. And, of course, bonus points if the products being produced will be long-standing and integrated into projects like construction.

Locking away carbon in long-term projects—like embedding CO_2 into construction materials—is a powerful weapon at our disposal in the fight to remedy our climate. Because it might sound surprising, but the cement and concrete industries cumulatively account for an astounding 7 percent of the entire world's carbon dioxide emissions.[73] The process is incredibly resource- and carbon-heavy, with concrete being used essentially every-where around the world.

Cement and concrete are often discussed synonymously, but they are two different materials. Cement is like the "glue" used to make concrete. For every ton of cement produced, about a ton of CO_2 is released into the atmosphere. But curing concrete with CO_2 instead of the water normally used can make the entire process less resource-intensive and become a nearly permanent housing for the carbon. CO_2 is essentially mineralized through conversion into calcium carbonate during processing, trapping it safely away without the typical concerns of it leaking from a subterranean storage site.

The carbon footprint associated with concrete manufacturing can be lowered, in conservative estimates, approximately 30 to 40 percent.[74] Some estimates have that number as high as 70 percent. Moreover, the new mix of minerals within the concrete can actually make it stronger, creating an equally enticing market incentive for using fewer resources in superior products. Manufacturing processes also can be substituted into existing facilities and require smaller amounts of time to cure (about 24 hours, compared to the typical one to four weeks).

Several companies have now pulled ahead in producing (and more importantly, patenting) parts of the process. Nova Scotian start-up Car-bonCure has already partnered with numerous concrete manufacturers. And the CarbonCure website even includes a tally for the amount of CO_2 they've sequestered (though they say they're ideally looking to lock up 500 megatons of CO_2 emissions per year).

New Jersey-based Solidia Technologies, which owns dozens of patents for their take on the process and uses AI to assist in speedier R&D phases, boasts savings of energy with kilns requiring less heat to produce the same product. They also estimate that if global production sites were to take up this resource-lighter process, nearly 3 trillion liters of fresh water could be spared every year. Further, Solidia estimates that the energy saved from a less resource-intensive process can save about 260 million oil barrels' worth of energy each year.[75]

CO_2Concrete LLC in Sante Fe, meanwhile, even cites its product as exceeding industry standards in strength.[76]

All things considered, utilizing CO_2 in concrete building materials as a "carbon sink" is estimated to eventually remove and store about 0.1 to 1.4 $GtCO_2$ per year.[77] That's another mere drop in the bucket of our estimated annual emissions—but when we do hit peak emissions, this will help us hit carbon neutrality faster on a global scale. And ideally, should we reach carbon neutrality without its help, assist with the arduous process of drawdown. Given that the process can mitigate the overall carbon foot-print of new buildings, the possibility of achieving a net-zero building can also be more attainable, likely spurring more firms to undertake them (if for no other reason than the tax credits). We have a unique chance right now, as we increasingly urbanize, to integrate these materials meaningfully.

A whole new branch of architecture is now aiming to store captured carbon within sustainable building materials. So-called "regenerative ar-chitecture" takes the carbon stored in materials like wood—in which trees have captured CO_2 during growth—and treats them for largescale proj-ects. And by some estimates, nearly 4 tonnes of CO_2 emissions can be avoided for each ton of dry wood used in place of concrete.[78]

For a construction project to be considered regenerative architecture, the carbon captured by building materials must exceed what is spent on building. Of course, integrating manufacturing-heavy items like PV solar panels can immediately tip you into the carbon positive, but the time-frame of "carbon payback" is where meaningful scales come into play. According to regenerative architects themselves, the real hero projects for carbon savings are retrofits, given that the concrete and steel of a build-ing's skeleton are already in place. Retrofitting buildings to save more energy through insulation, water efficiency, next-gen smart glass, or ad-vanced pro-energy building materials—along with stored carbon in wood—that's where projects can really do the most good.

Regardless, regenerative architects claim that by using prefabricated mass timber for structural components, new buildings can be constructed faster, with less waste and for less money. Moreover, engineered wood products can also be prefabricated and then assembled like large pieces of furniture, further reducing construction costs.

Naturally, building with mass timber conjures the imagery of devastating fires that razed old New York and old Chicago. But new advancements in engineered wood products (EWPs) and materials sciences have created new possibilities for using timber safely. Newer cross-laminated timber (CLT) and nail-laminated timber (NLT) consist of multiple layers of wood planks stacked at 90 degrees, either glued or nailed together under pressure. The material thickness within this composition actually increases resistance to burning.

Tests carried out by the United States Forest Service and the American Wood Council at the Bureau of Alcohol, Tobacco, Firearms and Explosives Fire Research Laboratory have demonstrated that EWP panels burn slowly when exposed to fire. This creates outer layers of char acting as a protective buffer zone, leading to typical mass timber materials exceeding building code requirements consistently.[79]

At the time of this writing, the world's tallest timber building is the 85 m (279 ft) tall Mjøstårnet building in Norway. Just how much CO_2 can a regenerative architectural project like that store? One series of residential units in the UK taking up about 3,852 m³ was estimated by engineers to have stored about 2,600 tonnes of net CO_2 (roughly 520 cars' worth of emissions for a year).[80]

According to a study published in *Nature*, one average five-story residential building structured in laminated timber can store up to 180 kilograms (396 pounds) of carbon per 10.8 square feet on average—roughly three times more than trees in a similarly sized patch of forest would store.[81] The study also noted that increased use of mass timber in construction worldwide could result in a range of between 10 million tonnes of reduced carbon per year on the very low end and 700 million on the high end.

Of course, for regenerative architecture to qualify as a reliable carbon storage tool, upstream and downstream CO_2 costs need to be calculated. Logging, manufacturing, and transport of materials are the upstream costs, whereas lifecycle and overall use of the building (including its eventual disposal) are the downstream costs. Supporters also ought to be aware that harvesting of timber plantations could leave little more than biological deserts behind, if unsustainably grown and processed. But for now, mass timber already has some impressive showings—a 70-story timber skyscraper in Tokyo, an 18-story timber building in Vancouver, a 200,000+ square foot university residence in Arkansas, and an incredible 80-story timber high-rise proposed for Chicago.

Another regenerative architectural possibility is hemp-infused concrete. Hempcrete, as builders are calling it, is a composite material in which the wood core of hemp plants has been mixed with lime and water into buildable Lego-like blocks. The blocks can then fasten together for quicker build times while also offering engineers a sustainable, carbon-negative alternative to help reach zero-emissions status. Airdrie-based Just BioFiber specializes in producing Hempcrete, and has reported the material helps control for air quality issues, heating and cooling retention, and is also fire-resistant. And research indicates they may be on to something: Japanese scientists at the University of Tokyo recently found that recycling concrete with the addition of discarded wood products can provide a benefit of increased tensile strength while improving the carbon storage of concrete used.[82]

Though it's a long shot, we also have new developments in turning CO_2 into carbon nanoproducts. These products, like graphene and carbon nanotubes, are highly valuable. And a few published papers tell us we might actually be able to refine waste CO_2 into the carbon nanomaterials used in advanced engineering materials—at scale.

Canadian start-up C2CNT (loosely: carbon dioxide to carbon nanotubes) claims they can do precisely that. The team begins by splitting CO_2 into its constituent components carbon and oxygen with a molten carbonate electrolysis process. The process is similar, in a way, to how aluminum is produced from its own oxide (aluminum oxide). Except in this case, carbon dioxide drives the process, and the team derives carbon nanomaterials instead.[83] Reportedly, powering the process with renewable energy and utilizing about 2 to 5 tonnes of CO_2 per day in processing.

If C2CNT's efforts are successful, the economics around carbon capture will have changed significantly. And fortunately, they may just be onto something: the group has published papers outlining their process in notable academic journal *American Chemical Society*. They have also officially been recognized as a finalist in the CarbonXPrize, an international competition centering on carbon use.

Crucially, though, aside from monetizing carbon dioxide itself, market forces will also play a role in monetizing carbon credits. More and more governments around the world appear to be penalizing emissions in some way or another. A Nobel Prize was tellingly awarded to an economist back in 2018 for working out pathways toward realistic carbon pricing—multiple signs indicate they are becoming more likely and desirable.

New developments in commercializing carbon off-set credits also continue to evolve. A company like San Francisco-based SilviaTerra utilizes Microsoft AI and satellite imagery data to literally count trees and assign carbon off-set values for plots of land through data-driven forest inventories. With their data in hand, landowners can essentially certify their sequestration data to plant trees for cash—while companies who wish to obtain that off-set can pay to fund it. The Seattle-based Nori carbon marketplace can help even landowners with just a few acres attempt to monetize their inventory.

It is important to note, as an aside, that some CO_2 utilization methods do not fully "remove" carbon from the atmosphere—especially not if we consider the entire span of a CO_2 molecule's voyage throughout the carbon cycle. If we are capturing carbon from the atmosphere, for example, and then simply reusing it in a fuel, then this is more of what the literature calls a cycling system. Realistic assessments must consider rebound or "leakage" effects in which CO_2 emissions may occur either upstream or downstream from its application. If we do, only a few utilization methods actually result in a net loss of CO_2, although the science is still very encouraging with natural sinks involved as well.

CARBON CROPS: THE BECCS PROMISE

One of the more defensible arguments against using hi-tech installations to draw down CO_2 is that they inherently involve a "carbon payback" period. This is the period in which the carbon later captured will need to make up for the emissions caused by the mining of resources for technology and the expenditures involved in construction and maintenance.

Developing a crop with superior carbon-capture capabilities, on the other hand, carries essentially only the up-front cost of research and development. Deploying one of the Chory lab's supercharged crops carries very little carbon footprint after the seed is developed. The moment you engineer the seed, your carbon payback period begins almost immediately and requires very little payback going forward. A strong argument can therefore be made for relying on biological methods entirely aside from afforestation options, which do not necessarily bring with them vast money-making opportunities to drive the process. This is where BECCS comes in.

To recap, BECCS stands for bio-energy with carbon capture and stor-age. The basic premise is that we can grow fields full of crops with renew-ably powered equipment. The crops draw in carbon as they grow. We then harvest those crops sustainably and burn or convert them into energy forms as we capture the carbon produced in the process. The result is a net loss of CO_2 from the atmosphere and fuel for us to commercialize.

Sounds like a no-brainer, right?

In a way, it does, but the BECCS option has also been criticized for a number of reasons. The UN's climate panel recently warned against the prospect of creating "vast bioenergy plantations" for the sake of fighting climate change.[84] They (and others) have warned against the possibility that water requirements for growing all our new fuel crops will skyrocket. Farmers may also see more commercial incentive to grow fuel crops over food crops, raising the price of food for various communities worldwide. Particularly so if this occurs on a very local level, in which a town's only farmers cash in on energy crops instead of food crops. There is also the possibility of biodiversity collapse should farmers pursue monocultured fuel crops, as well as the dangers of industrialized agriculture in fertilizer use and nitrogen problems.

Worse, by some estimates a BECCS-heavy portfolio would require an area of landmass roughly the size of India to make a significant impact.[85] That implies many, many square miles of carbon crop fields to scale this option meaningfully. In fact, some researchers estimate that based on previous BECCS plantations, hundreds of thousands of them could be necessary overall to make a significant contribution to greenhouse gas reductions.[86]

And if biofuels are to be established on ecosystems which need to be transformed beforehand in any way, then we will have that aforemen-tioned carbon debt to pay off. Analysts have warned that "first generation" biofuel crops must operate for years before settling such a carbon debt. In the case of some sugarcane-based ethanol systems, the carbon debt has been as high as 17 years. In one case, a poorly planned biofuel installation in Brazil replacing tropical forest on peat-lands with a palm-based oil biodiesel system tallied itself a carbon debt of roughly 840 years.[87] But that, of course, was horrendously plotted and in no way defines many reasonable biofuel plantation proposals.

On the other hand, the biomass used in BECCS projects need not be carbon crops grown purely for greenhouse gas reduction. These materials could include agricultural and organic industrial waste as well. Even if that might be logistically challenging, the potential for some carbon off-set exists.

In terms of the role that BECCS operations can play in our global cli-mate response portfolio, some researchers believe that with all kinds of potential crops up for potential use, net carbon removal could vary any-where from 3–20 $GtCO_2$ per year.[88] Of course, some of those estimates are based on simply keeping carbon from entering the atmosphere in the first place. But if an end result for BECCs applications looks anywhere near even a mid-range of that estimate, that is a considerable weapon in our global portfolio. Estimates on the upper end of that spectrum envision massive land usage going along with these operations, along with strict policies in support (and are admittedly rather optimistic, as a result).

Fortunately, one BECCS proposal may help to bypass the land use criticism. The concept of using algal biomass to produce bioenergy can allow engineers to tap the vast expanse of marine farming areas. As such, algal bio-energy carbon capture and storage (ABECCS) may be superior in cases where land space is becoming a concern—particularly if strains of algae bound for ABECCS have been bioengineered for superior bio-energy traits.

Ultimately, BECCS proposals will seem more reasonable if producers can generate revenue for themselves with crops or algal strains able to accomplish additional feats besides merely storing CO_2—while staying non-competitive with agriculture. Research indicates this is a likely reality of future biofuel crops, so that's certainly promising.

SOILWORK:
INCREASING SOIL STORAGE

Terrestrial carbon storage options like afforestation and the Chory lab's supercharged plants clearly have potential to remove CO_2 and help to reverse a large driving force of climate change. But aside from how we absorb CO_2 from the air and the mechanisms for storing it, there still remain several options for improving the medium in which carbon is to be stored—the soil.

Naturally, there can be better or worse soil conditions for storing car-bon. And as it turns out, soils will hold and store much more atmospheric

carbon if they are high in organic matter. This is an outcome we can encourage through sustainable farming practices that more closely emulate
natural processes found throughout the environment as opposed to the
kinds of agricultural processes we currently use. Fortunately, scientists are
hard at work investigating methods for improving our soils to generate
outcomes of superior carbon storage.

By some estimates, just the top five to six feet of the Earth's soil holds
several times the amount of carbon currently in the atmosphere.[89] In fact,
according to soil scientist Asmeret Asefaw Berhe—at a TEDxTalk in
which she implores us to stop treating soil "like dirt"—soil stores around
2,500 gigatons of carbon (roughly more than three times the amount of
carbon in the atmosphere and four times the amount stored in all living
plants and animals).[90] And merely replenishing and protecting the world's
soil carbon that we have lost over time would off-set up to 5.5 billion
tonnes of CO2 every year—just under the emissions total coming from
the United States.[91] Scientists therefore believe the soil can hold much
more carbon than it currently does, presenting us with another powerful
solution for drawing down CO_2.

Soil conditions like peatlands, for example, are incredible sources of
carbon storage. Peat (or "turf," depending on where you happen to live) is
what we call soil when it is full of decayed vegetation and other organic
matter. Peatlands are found in bogs, mires, and moors where the earth is
often black and rich in texture. Although peatlands only cover about 3
percent of the world's land surface, they store twice as much carbon as all
of the Earth's standing forests.[92] Conserving peatlands can therefore be a
considerable source of carbon storage over time.

However, soil carbon sequestration can also be incredibly variable. It is
highly dependent upon local vegetation, the underlying geology, and local
climate conditions. As a carbon sink, soil sequestration is also slow to
accumulate. So we cannot count on immediate, high-impact carbon removal here. But we can begin the work of accumulating those savings
right now, piecemeal as they may be to start.

Scientists typically speak about improving soil carbon storage through
a blanket term of *regenerative agriculture*. Regenerative agriculture includes a number of pro-carbon farming practices designed to maximize
the health of soil, leading to improved crop growth and carbon retention.
Practices like incorporating more compost and cover-crops into fields,
diversifying and rotating crops, reducing chemical inputs, incorporating

tree conservation into agriculture (agroforestry), avoiding cattle overgrazing, excessive tilling and plowing of land, and more. In all, there are more than 20 techniques that fall under the regenerative agriculture umbrella. As a result, farmers can play a very significant role in our efforts to convert more atmospheric carbon into soil carbon.

Cover crops imply planting a secondary crop to cover the soil where it is not being used by the primary crop one intends to harvest. In between stalks of corn, for example, one might plant a cover crop of clovers. Rather than exposing parched soil to bake in the sun, leafy coverage can save soils below from moisture loss, oxygen-driven carbon loss, and erosion. And plant nutrients can be shuttled down to assist an assortment of soil biome supporters.

Reducing "tillage" means not churning up the topsoil to plant new seeds. Doing so exposes soil content to oxygen. Fortunately, new technologies discussed in the chapter about ag-tech will enable farmers to avoid excess tilling.

In reality, integrating regenerative farming practices means potentially planting two or more crops in the same field and harvesting them at different times. It means integrating nitrogen-fixing plants to help avoid synthetic versions, and possibly integrating animal species into cropland areas to reduce overgrazing and stockpiling natural fertilizer further.

Project Drawdown has actually listed "silvopasture" as its ninth-overall climate-remedying solution for managing our carbon, which would place it above options like solar power and electric vehicles. Silvopasture is an ancient form of farming that includes integrating animals and trees into agricultural spaces. Livestock animals are essentially moved between rotating enclosures, ensuring they do not overgraze one area and that they do spread helpful natural fertilizer around. Drawdown has estimated that if farmers could increase silvopasture practices from today's 351 million acres to 554 million by the year 2050, we can reduce CO_2 emissions by about 31 gigatons.[93]

Projecting the amount of carbon able to be sequestered by soils around the Earth can be a fairly speculative process. In academic literature, one might find very optimistic scenarios in which over 10 percent of anthropogenic CO_2 emissions can be stored every year.[94] Other times, this estimate has been as low as about 2 to 3 percent. To settle this discrepancy, I often look to the most and least optimistic organizations pledging to help restore soil carbon for any evidence of a moderate take. In a 2018 interview with the organization Soil4Climate, Rattan Lal of the Carbon Management and Sequestration Center at Ohio State University stated that regenerative landscapes could potentially restore up to 150 Gt of carbon in the world's soil in 80 years.[95] This seems to be a mid-range estimate, depending on when we begin the benchmark of measuring the percentage of our emissions (now with high emissions, or over the next decade with a smaller amount).

The option of using soil sequestration as a strategy for drawing down CO_2 is a comparatively cheap one. Direct air capture technology, you will remember, captures carbon for about $100 per ton at this time of writing. That number could certainly change, given the exciting new technologies also being developed and potentially able to scale up in the future. But the economics of simply paying farmers to initiate these strategies can be cheaper. Boston-based agricultural start-up Indigo Ag, for example, has launched a "carbon marketplace" in which farmers can be paid $15 per ton to capture more CO_2 in the soil. That is far cheaper, but the trade-off is in the timescale of achieving these results. Direct air capture begins immediately; soil takes time.

Several notable regenerative agriculture campaigns are ongoing at the time of this writing. The Terraton [sic] Initiative, for example, is a global effort aiming to remove one trillion tonnes of CO_2 from the atmosphere and store it in the soil. The campaign consists of challenges to encourage technological innovations, a championship cup to foster healthy competition, a real-time network of experimental data for open source improvements, and a "carbon market" for providing monetary incentives to farmers. The "4 per 1000" initiative, launched in 2015 at COP 21, calls for increasing the carbon storage of agricultural soils annually by 0.4 percent within the first 30–40 cm.

Aside from encouraging more soil sequestration of CO_2, projects like Terraton should yield invaluable data for how farmers might maximize carbon storage going forward. The answers aren't always as obvious as they might seem. One study by researchers at the University of California Davis, for example, found the use of compost and cover crops boosted soil carbon content by 12.6 percent, while only using cover crops alone created a significant *loss* of net soil carbon.[96] The good news, though, was that the study also found a combination of compost and cover crops can add soil carbon by 0.7 percent; an improvement on the 4 per 1000 challenge.

Ultimately, soil farmers offer us a compelling opportunity to finance carbon storage at a very affordable rate (comparatively speaking). But we will need to act quickly if we want to cash in on soil's potential for storing carbon. Recent research tells us that warmer soils do not store carbon as well, compounding that urgency. If soils all around the world warm further, researchers have estimated that increased microbial activity therein could result in a 9 percent increase in atmospheric CO_2 as a result[97]—so we need to get moving.

Another application of agricultural technology is the concept of "perennializing" crops to essentially avoid tillage altogether. Most of our agriculturally important crops are annuals that undergo a life-cycle true to their name—they must be re-planted every year. This continual re-planting is energy-intensive and often includes agriculture practices that incur heavy soil-carbon losses as seeds are planted.

Perennial plants, on the other hand, are species that live for more than two years and typically include woody shrubs and trees. We do have a few perennial crops in our agricultural sector, but they account for only about 13 percent of the world's food production in surface area.[98] Perennials offer more extensive root systems, enhance soil biodiversity, make more water available to other plants, and often capture and sequester more carbon.[99] Expanding the list of perennial crops in our agricultural staples can therefore ultimately increase soil carbon storage.

The prospect of "perennializing" more of our agriculture can, in one sense, refer to using more perennial crops in general. Switching to naturally perennial crops is one way to achieve it. However, some scientists are also working on introducing perennial growth cycle traits into traditionally annual plants through genome editing. Researchers are currently limited in this area, both by underfunded research and lack of popular support for that level of genetic modification on foods we eat, but the work continues.

The potential for increasing soil carbon storage simply by growing perennial versions of staple crops may encourage more research in this area. Rice, for example, is considered an annual crop but has perennial strains in tropical areas. These traits could potentially be hybridized into more widely used variants in other parts of the world, creating a powerful impact on soil carbon maintenance everywhere rice grows. Progress in developing perennialized rice has more recently been made in China.[100] And a new perennialized grain called Kernza is now destined for cereals and beers.[101] (Kernza is a wheat we'll undoubtedly hear much more about in the near future: US food giant General Mills has invested heavily in developing new cereals with it. Lab technicians are now, at this time of writing, working on different Kernza strains for different uses like larger yields and smaller dwarf versions for bread varieties.)

We now have a fairly solid understanding of what improves the carbon storage potential for good soils, and we do not need to rely on expensive equipment or untested materials to capture CO_2. This makes soil management a great alternative for other carbon capture methods, given that it includes other benefits for crops.

Project Drawdown estimates that today's current 108 million acres of regenerative fields can potentially scale to 1 billion acres by the year 2050.[102]

SCATTERING THE SILVER

Fortunately, thanks to recent technological innovations, the options for CO_2 removal are increasing in number and falling in cost. But we've certainly got our work cut out for us. Our current carbon level at the time of this writing, 415ppm, is the highest concentration of atmospheric carbon dioxide our planet has had since the Pliocene.[103] There were trees at the South Pole back then.[104]

If we truly do become serious about fixing the climate crisis, we'll begin by lowering emissions as quickly as possible. At the same time, we will be developing NETs to cut out some of the damage while we transition. And then, as we approach and reach net zero emissions, we will continue developing and deploying NETs in concert to achieve drawdown.

If we are using a combination of all the tools at our disposal, we will be stopping deforestation as we initiate more and more reforestation and afforestation campaigns. Even now, without the technological bells and whistles, scientists estimate that "all natural" climate solutions of afforestation and land management can easily account for more than a third of the mitigation efforts we need by 2030.[105] We will potentially deploy, in earnest and with grace, supercharged plants and crops in select areas to maximize carbon storage and speed up the process. We will have artificial photosynthesis equipment reducing atmospheric CO_2 as we simultaneously produce fuels for hard-to-decarbonize sectors.

We will increasingly turn to hybrid tech like the biosolar leaf for carbon negative sources of consumer goods, because receiving help from self-sustaining organisms is a very compelling concept. After all, organisms can reproduce rapidly. They have a higher photosynthetic conversion rate over plants and trees. They can remediate environmentally damaged areas. They can simultaneously produce additive products. They can be enhanced more readily through bioengineering. They do not compete with other land use options like farming. They can be grown in bioreactors, if needed. Microorganismal carbon capture may well become a significant component of our silver buckshot.

We will be trapping carbon in products like concrete and smart buildings. A mass roll-out of direct capture machines is certainly not unlikely. We will, I suspect, be partially subsidizing direct air capture equipment, provided it is renewably powered. By one estimate from researchers recently published in *Nature*, by the year 2100, direct air capture machines alone could be using a full quarter of global energy[106]—illustrating how the need for immediate action may compound with the years.

The use of NETs will certainly impact our planet's carbon cycle in the future. The same kinds of buffers we enjoy in mitigating how much CO_2 we pump into the atmosphere now, like the oceans and soils sponging up excess carbon, could respond similarly in our removing that CO_2. The amount of CO_2 absorbed by plant life may decrease when atmospheric concentrations are lower, slowing the rate at which we can fix this problem. Oceans will also provide less of a sink with smaller amounts of atmospheric concentration. Worse, research indicates the oceans could "rebound" in releasing more CO_2 into the atmosphere as it loses the CO_2 we take.[107] This compounds our need to energize the NETs at our disposal immediately to solve these challenges while we still have time.

With everything from this chapter in mind, then, it is certainly worth repeating that it is far easier to keep CO_2 from entering the atmosphere than it is to remove it.

CHAPTER 10

DAMAGE CONTROL

■

> We can't always predict when our meddling will save
> species or when it will backfire. But we had better try.
> We have pulled a few species back from the brink—the
> California condor, the Whooping crane—by insinuating
> ourselves in their lives as puppet mothers and migration
> guides, so intimately that I squirm at their lost dignity and
> wildness. But then I remind myself: that dignity trip is my
> baggage, not theirs. They just want to live.
>
> —EMMA MARRIS, *HUMILITY IN THE ANTHROPOCENE* (2015)[1]

By the year 2050, it has been projected that we will have more plastic in our oceans than *fish*.[2]

Creating a more livable planet is certainly about taking steps to remedy climate change and transitioning toward a more sustainable way of life. But it is also about taking responsibility for the systemic planetary damage we've already inflicted. It is about ensuring that the many species impacted by climate change—including all members of our own species—can be spared the worst effects of our environmental neglect. We must take a step back from post-industrial largesse to survey the costs from times in which we were less aware of our impact on the planet. We simply must care that our oceans are full of garbage and plastic and endeavor to do something about it.

It's true that our ability to create a material like plastic has been an incredible feat for us. It was truly a remarkable accomplishment when, in 1869, the world's first semi-synthetic polymer (celluloid, the first industrial plastic) was produced by John Wesley Hyatt—inspired by a New York firm's $10,000 offer for anyone who might find a viable substitute for ivory. The cotton-derived cellulose Hyatt used could be crafted into different shapes, essentially imitating natural substances that manufacturers had been hunting to depletion until then. In a case of historical irony, the discovery was touted at the time as a saving grace for turtles—given that the substitute also worked for tortoiseshell imitation. Ironic, of course, because we now we see turtles getting stuck in plastic six-pack holders and often perishing.

The problem is that plastic materials have proven so useful that our world is now inundated with them. Researchers analyzing the global output of plastic over the past few decades have released some startling statistics to reflect this. They estimate that with all the plastic we send to landfills, the material somehow reaches our oceans at a rate of roughly a garbage truck's worth of contents every minute. Scientists further estimate that the ocean is already filled with more than 165 million tonnes of plastic.[3]

Worse yet, the development of "microplastics" has created an even larger problem for the world's oceans. Microplastics are a versatile material in the plastic family, but are also so small they are often simply consumed by unsuspecting marine life when they reach bodies of water—marine life we then purchase at local markets for our own consumption.

Aside from ingesting plastic through marine life, we have also tainted many of our drinking water supplies with these materials. In the United States alone, the World Wildlife Fund has reported that roughly 94.4 percent of tap water samples contain at least some remnants of plastic fibers.[4] Indeed, according to the same assessment, you may be ingesting about five grams of plastic every week. That is roughly the equivalent of eating a credit card—per week. Another recent study found that humans may eat, on average, more than 50,000 pieces of microplastics every year (a figure which rises if we consider airborne, inhaled microplastics).[5]

Yet, undoing our damage is about more than cleaning up our plastics. Increasing global temperatures have disrupted ecosystems everywhere we dare to look, causing mass migrations of animal life into cooler areas. Some species are lucky enough to simply be displaced, migrating by foot into safer pastures. Sadly, however, some species cannot outrun the hotter temperatures—because while seed dispersal will relocate trees and plant life into more habitable areas, they cannot simply get up and move away.

The good news is that scientists around the world are actively working on solutions for these problems. Some things will be easier to fix than others, but a number of promising new proposals are in development to achieve the following outcomes::

- Assisting populations dealing with climate change until the climate becomes less irregular.
- Remediating the damage of synthetic materials in the environment.
- Restoring pre-industrial biodiversity.
- Removing pollution from the air, ocean, and terrestrial sites.

These problems are not insurmountable. Just as we should be able to stabilize Earth's carbon cycle, we can certainly remove much of the plastic from our landfill sites and drinking supplies. We can assist populations (including our own) struggling to survive with an erratic climate full of extreme weather events. We can attempt to make the oceans less acidic. We can promote biodiversity to stop the unprecedented—and accelerating[6]—levels of mass animal extinctions ongoing worldwide. We can finally live up to our responsibility as the planet's most sentient and capable species to practice stewardship of the environment in responsible directions. And we can do it with new science and new technology.

DEALING WITH PLASTICS

I'll admit the problem with plastic is actually worse than I implied.

For years, companies have used "microbeads" of millimeter-sized plastics in personal care products, which we often then unwittingly washed down our drains. From our drains, these microplastic materials made their way into bodies of water and subsequently into the fish we eat. That was where I left off, but plastic chemicals have circulated through our water systems so thoroughly that we now find them essentially everywhere on Earth. In the Arctic, for example, scientists have found plastics inside the eggs of seabirds more than 100 miles away from the nearest human settlement. Sadly, that means plastic-derived contaminants are being ingested by birds and then maternally transferred to eggs, according to researchers.[7] This suggests the saturation level of plastic waste is shockingly high even in remote locations.

One problem is that scientists aren't entirely sure how to measure animals' plastic levels reliably and continuously enough to assess the true scope of the problem. We once thought we knew how much bisphenol A (BPA) that humans were being exposed to. But researchers have more recently found that our own exposure to this particular plastic is far higher than previously estimated. Human exposure to BPA has been underestimated by as much as 44 times.[8] That's quite a margin of error.

I could certainly go on. It gets even worse. The more you research, the more you see terrible headlines describing what plastics are actually doing to our environment. Headlines about a whale being pulled from the ocean with 88 pounds of plastic bags in its stomach, for example, or a two-foot plastic shower hose expunged from a dolphin's throat, or heartbreaking images from a show like *Blue Planet* of a mother albatross unwittingly feeding plastic bits of garbage to her young. Crushing, never-ending images like these compound the urgency of addressing our plastic issues.

Scientists and engineers have split "the plastic problem" into multiple categories. Some scientists believe that we can accumulate and remove much of the trash from our oceans. Others target stopping the flow of plastic in rivers long before it has a chance to reach the ocean. Research certainly indicates that rivers are a powerful place to start: the flow of trash from just ten rivers worldwide reportedly contributes about 90 percent of the plastic that will eventually end up in our oceans.[9]

One non-profit, The Ocean Cleanup, has designed devices to work both with ocean and river plastic removal. For example, Ocean Cleanup's team has created a solar-powered robot called The Interceptor equipped to catch surface-level plastic floating downstream in rivers. An attached conveyor system then feeds all the plastic and garbage into separate compartments for sorting and dispersal.[10]

The Interceptor is anchored to a riverbed, away from boats that may pass by. The equipment is then strategically positioned over a part of the river to catch plastic and trash floating by, with another unit placed further downstream. An autonomous and solar-powered system then separates these materials, and will alert local operators when a collecting dumpster is full and ready for pickup. Over the course of a typical day, engineers claim, the device can extract as much as 50,000 kilograms of trash (depending on the river). There are currently Interceptors operating in Indonesia, Malaysia, and Vietnam. Others are scheduled for deployment in the US.

There are also devices made to target smaller, less accessible particles of plastic. A Biomimicry Global Design Challenge top 10 finalist, the "Floating Coconet" mimics the biological strategies of manta rays and basking sharks to filter food from passing water.[11] The team has designed a filtration system for trapping tiny plastic bits passing through streams, thanks to one of nature's designs for trapping small particulate matter.

Or consider a winning project from Google's 2019 Science Fair. Eighteen-year-old Fionn Ferreira from Ireland took a concept inspired by NASA and used magnetic liquid to attract microplastics potentially floating in pools of water. So-called "ferrofluid" liquids were invented by NASA engineer Steve Papell back in 1963, originally intended to magnetically direct rocket fuel toward inlet pumps in weightless environments. Young Ferreira took the concept and applied the nonpolar nature of plastics and oils to deduce a connection for assembling them in liquid. At Google's Science Fair, Ferreira was eventually able to capture about 88 percent of the microplastics in water samples taken on display.[12] The concept could easily apply to wastewater treatment plants, which sometimes do not even screen for microplastics at present.

One Dutch innovation has been using underwater bubble generators to get submersed plastic moving along the river to rise to the surface. Once at the surface, trash is collected by the riverside. The device itself is a long, perforated tube extending across the bottom of canals with air being pumped to shoot through the tube's holes. The Bubble Barrier, as it is called, allows fish to swim by essentially unaffected while pushing plastic materials upward.[13]

Stopping the flow of plastics into our oceans will likely require an approach with multiple strategies. To start, we can employ a suite of "catch" strategies at wastewater plants and rivers to stop the vast circulation of plastic from continuing unabated. One recent assessment from Washington State University estimated that so many microplastics move through the American water supply that roughly eight trillion pieces will pass through wastewater treatment plants every day before moving on to other aquatic environments—providing us plenty of opportunity to mitigate part of the plastic crisis through filtering at waste streams.[14] Transitioning to truly biodegradable bioplastics or finding ways of using less plastics will also help immensely. For the plastics already in our oceans, however, the problem will be much more challenging.

Once in the ocean, plastics are much harder to seek out and capture.

Fortunately, Ocean Cleanup has reported some success in this arena. The organization tapped the engineering minds of Dutch scientists for an enormous floating arm device that essentially reaches out and gathers floating plastics for automated capture. The team has reported capturing and removing plastic from the Great Pacific Garbage Patch—which I recommend Googling, by the way, to see what they're up against.[15]

Incredibly, Ocean Cleanup's equipment can catch garbage and plastics ranging from the size of cartons and containers to fishing gear and even microplastics. The device's arm extends outward into a U shape with a net dropping below the surface, catching objects flowing with the current. Fish can reportedly swim beneath the device, making it relatively safe for marine life. The team aimed to launch a full-scale cleanup network throughout 2020 targeting the Pacific garbage patch specifically (but may have understandably been delayed). After that, the team will move on to several other well-known current-dragged patches of trash.

Some researchers have combined new advancements in aerial reconnaissance with projects of garbage tracking—and have mapped where our trash ends up by utilizing drone swarms and mounted cameras. A specialized AI software program can then assess the images to recognize trash like bottle tops and plastic bags from objects like shells and jellyfish. With this data, scientists hope to understand where our oceanic trash lifecycle starts and ends, mapping out hotspots and directing future relief efforts.

Or consider the Roomba-like WasteShark drone from RanMarine, another Dutch effort. The aquatic drone can operate on autonomous mode or while being piloted remotely by an operator. The device essentially roams coastlines, where it seeks out trash floating on the water's surface. It is able to carry up to 352 pounds of trash and can operate on its own for about 16 hours.[16]

WasteShark can also gather air and water quality data along its route, while filtering dangerous chemicals like oil, arsenic, and other heavy metals. The system comes with an on-board laser image detection system capable of spotting larger objects in its path and avoiding a collision. Some hotels and other businesses with harbor fronts inundated by drifting trash have hired RanMarine to regularly dispatch its aquatic Roomba-like tech to gorge on floating refuse. The company has explained via TEDxTalk lecture that the drones don't harm fish, which generally swim away from it.

Another start-up close to Dubai, BluePhin Technologies, has marketed a similar device. BluePhin's aquatic Roomba-like machine is AI-powered and also seeks out trash, collecting up to 370 half-liter bottles of waste in just a few hours. An on-board sensor dispatches the bot back to land for harvest when the trash collector is full.

The EU, meanwhile, has launched Project CLAIM: Cleaning Litter by Developing and Applying Innovative Methods in European Seas. The effort has deployed five new marine cleaning technologies with an aim toward improving European waterways and bodies of water. Tech 1 gathers microplastics at wastewater plants. Tech 2 breaks down microplastics by blasting them with UV radiation using a unique photocatalyst to speed up the degradation. Tech 3 is another river-based filter system for catching plastic that might be drifting downstream. Tech 4 is a small-scale pyrolizer system for turning plastic debris back into fuel—more on this concept soon. Tech 5 models underwater data to produce maps of micro and macro litter concentrations for better upstream intervention.

Sadly, some research indicates that the plastic we see out in the oceans may actually just be the tip of the iceberg. Or, according to oceanographer Erik Van Sebille from Utrecht University in the Netherlands, "less than the tip of the iceberg, maybe a half of 1 percent of the total."[17]

Some speculate that the remaining 99 percent of our plastic ends up in the deepest parts of the ocean, buried in sediment or potentially fragmenting into pieces so small we can barely detect them.[18] Once plastic enters the ocean, it becomes weathered and more likely to dissolve into smaller substances. A combination of UV radiation and abrasion from the ocean's water can disintegrate larger pieces of plastic into microplastics and even "nanoplastics." And because nanoplastics are so small, they can accumulate in bloodstreams and cross the blood-brain barrier in fish—raising the grim prospect that the same phenomenon may also happen to us.[19]

Research teams aren't going down without a fight, of course. An inventive team of EU scientists has co-opted the biological properties of jellyfish in the hopes of capturing microplastics.

As it turns out, jellyfish mucus can efficiently capture nanoparticles.[20] With that in mind, one of the projects funded by the EU's H2020—an R&D program funding initiatives that aim to solve societal challenges—was a project called GoJelly. The program's team designed "a gelatinous solution for plastic pollution" by adapting the slime as a biofilter for water treatment areas to catch microplastics. As of this writing, the team is still working on assembling a synthetic version of the mucus to deploy at a mass scale for use anywhere we want to screen for microplastics. Someday synthetic jellyfish bots may roam the oceans, sponging up microplastics and returning to a harvest zone for collection and subsequent re-dispersal. The GoJelly program will run until December of 2021, when the results of the campaign should be much clearer.

On a positive note, jellyfish are actually older than dinosaurs and have lived through five of Earth's mass extinction events—including the "Great Dying" which wiped out up to 96 percent of all marine species on the planet. So, I wouldn't worry as much about natural jellyfish struggling with the ocean's microplastics.

PURGING PLASTICS:
DEGRADATION AND RECYCLING OPTIONS

Even if we develop reliable methods for collecting all the random plastic from our rivers and oceans, the question then becomes: what are we going to do with it?

For starters, the plastic needs to be separated from other bits of trash and gathered together—a process that has undergone rapid evolution over the past few years. In fact, we have a variety of techs now designed to accomplish this, increasingly automated and renewably powered. But whether plastic is separated from other trash at its disposal site or when it reaches a sorting facility, that's the first question.

On-site disposal separation is now increasingly driven by AI that takes advantage of new developments in object-recognition software and camera tech. CleanRobotics, for example, touts its TrashBot as one of the world's premier "smart trash cans" to analyze what is being deposited and sorting the contents accordingly. The company has integrated their tech into high-waste areas in heavily trafficked sites like airports, malls, and large-scale office buildings. TrashBot was famously a finalist for the $5 million IBM Watson AI XPRIZE targeting new international innovations in artificial intelligence.

Other attempts to integrate AI into the trash-sorting process have focused on consumer education programs right at disposal areas, using visual ID software to educate individuals about where and how to dispose of items. Take, for instance, the "Oscar" software, which analyzes trash you're about to throw out and notifies you of where to place certain components.

New advancements in machine-learning software have allowed computers to assess the visual cues of specific objects to determine what you're throwing out (and to educate you on where to place trash appropriately).

The use of AI-sorting may add a crucial component to motivating businesses to recycle if a verifiable, quantifiable degree of recycling one day results in subsidy credits. One can hope.

Some countries have utilized "reverse vending machines" where consumers input plastic waste to receive discounts on future purchases. The programs aim to help save sorting plants the Sisyphean task of categorizing various plastics. And, of course, to improve the decentralization of sorting and recovery efforts.

Other technologies use touch-based sensory analysis for sorting out recyclable materials. Researchers from MIT and Yale, for example, crafted the aptly titled RoCycle system to separate materials with an 85 percent accuracy rate while stationary, 63 percent of the time on a conveyor belt.[21] On-board sensors first gauge an item's shape, then a Teflon-tipped gripping appendage will test various tactile qualities to judge what an item might be. The gripping apparatus uses a "strain sensor" to estimate an object's size before assessing pressure points while grasping. Data from these metrics are collected and calibrated along with information about conductivity to assess metallic components of objects gripped. Machine-learning algorithms then assist the program going forward.

The Cortex Robot from AMP Robotics uses a combination of optical software to sort the objects entering a facility via conveyor belt and then tactile information to ascertain what must be done with objects, like the MIT process.

For decades, Western countries have responded to their recycling needs by shipping trash elsewhere for sorting—often, to China. But with China no longer taking the world's garbage, most others are now refusing as well. Philippines President Rodrigo Duterte even threatened a declaration of war against Canada if the country did not take back garbage it had shipped there.[22]

What some commentators suspect may happen is a combination of all the above approaches: a swipe-card for throwaway trash to help motivate recycling combined with robotic sorting tech, an assessment of one's recycling efforts, and subsequent rewards or punishments.

After sorting, though, we will still need to actually *do* something with plastic and garbage.

Fortunately, new advancements in materials sciences and bioengineering can help us. Scientists have now developed a powerful new series of microbes (or "tailored marine consortia" as they were fancifully labeled in research) capable of eating plastics destined for landfill trash digesters.[23] One team had actually found naturally occurring microbes doing some of the work, raising a small amount of hope that if we simply stop saturating the oceans with plastics, there are natural solutions. Until then, however, engineered variants of plastic-eating microbes may prove useful at recycling areas.

Bio-based plastic solutions often focus on the ability of microbes to break plastic down, or on the inherent biodegradability of bioplastics themselves. But that's something of a half-truth: many biodegradable plastics should come with an asterisk that reads, "biodegradable in certain conditions"—those certain conditions being the hot, wet temperatures of a treatment plant where controlled biodegradation can occur. If you were to simply toss an average biodegradable product into your garden, it likely wouldn't degrade for quite a long time.

Enterprising scientists working with the French company Carbios have engineered a unique solution: the introduction of "enzymated" plastics to create a product that self-destructs (read: self-degrades) over a set lifecycle of time. The team introduces enzymes into the plastic's material itself in the form of tiny pellets, unlike others that wait for plastic-eating microbes to find the product in the environment or treatment plants. According to one of their PR posts, the natural lifecycle of a Carbios plastic can be roughly 125 days for an average plastic bottle.[24]

The Carbios team, like many others in this book, originally began looking for natural analogues in the environment. They went to landfills. They went to plastic dumping sites. They studied the natural microorganisms breaking down plastics with the help of enzymes. The team studied the composition of those enzymes and engineered them to work much faster. As a result, the team not only intends to deploy the materials into bottles, but also to utilize plastic-eating microbes for on-the-spot biorecycling.

Normally, recycling plastic will degrade the material, but the de-polymerization process at Carbios essentially purifies the material, creating what they colloquially refer to as "virgin polymers" able to be processed an infinite number of times.[25] The team will officially launch their first full-scale, commercial-sized plant in 2021.[26]

Carbios was recently featured in *Nature* for their efforts in isolating a record-setting rate of plastic recovery. Along with the University of Tou-louse in France, the group published peer-reviewed research indicating that they could recover roughly 90 percent of plastic from an average water bottle—up from just 30 percent in today's plastic recycling plants. The team took a naturally occurring enzyme in the landfill-loving bacteria they had discovered, generated hundreds of mutant variants, and eventu-ally produced an enzyme 10,000 times more efficient than what they started with.[27]

Another team at Pleasanton, California-based Kiverdi uses a "reverse plastics" solution quite different from the Carbios team to start: gasifica-tion. Kiverdi breaks plastic apart into its fundamental hydrogen and car-bon components. Gases are essentially bubbled through a liquid where biocatalysts convert the gases into biodegradable materials. The results, they say, are biodegradable polymers able to be reused an infinite number of times.[28]

A variety of enterprising business models have developed around using plastic waste. Los Angeles-based ByFusion boasts turning 100 percent of its recycled plastic into concrete-like building blocks they call ByBlocks, assembled by their Blocker equipment. The device processes plastic through a combination of steam and compression to reshape plastics into bricks able to be used for construction (if you're picturing Disney's WALL-E compressing little blocks of trash, the end product actually looks a little similar—except much cleaner).

Fellow LA-based plastic recovery company TechniSoil Industrial simi-larly creates "plastic asphalt" by mixing plastic into ground-up blacktop. The TechniSoil team uses recycled PET plastics to replace bitumen, a sludgy, semi-solid form of petroleum commonly used to hold asphalt to-gether. Incredibly, the team uses a train of processing units for recon-structing the top surface of broken roads: first break up the top few inches of a street surface, move that material into a second unit that crushes the asphalt into uniform size, then move that into a mixer with liquid plastic. The result, they say, is a road that will last 13 times longer than a standard asphalt road, with a 90 percent reduction in carbon emissions (from all the additional trucks and resources used otherwise).[29]

Another strategy is pyrolysis, a process of thermally degrading materials in the absence of oxygen. According to research from scientists at Purdue University, polypropylene pyrolizers can actually turn plastic back into its native form: oil. Because many plastics are made from petroleum, a hydrocarbon, they can also be converted back to liquid fuel. And in this case, the research team found that the conversion process could actually be net-energy positive and even have a lower greenhouse gas emissions profile than mere incineration or mechanical recycling.[30] The oil derived from propylene, which makes up about 23 percent of total plastic waste, was proposed to be used in gasoline blendstocks.

This brings me to probably my favorite option: turning plastic refuse into jet fuel. A variety of companies and research teams have now demonstrated that re-forming plastic in specific conditions will yield fuel capable of use in jets. One research team led by Washington State University scientists discovered a method for turning daily plastic waste products into jet fuel by mulching it up and superheating it with a specialty catalyst. The team melted plastics at very high temperatures of 430 to 571°C (806 to 1,060°F) along with an "activated carbon," which is a processed carbon with increased surface area. The team was able to take an assortment of plastic waste products like milk bottles, plastic bags, and water bottles and ground them into three-millimeter chunks (roughly the size of a grain of rice). The carbon catalyst is added to the high-heat reaction to break down chemical bonds, then separated out to be reused again. In all, the team was able to yield a mix of 85 percent jet and 15 percent diesel fuel from plastic waste.[31] The process was cited to be easily scalable, pertinent for large-scale facilities or even farms where industrious do-it-yourselfers may want to create their own homemade diesel.[32]

In another compelling first (though not yet commercialized), researchers at Nanyang Technological University in Singapore were able to break down non-biodegradable plastics like polyethylene into usable fuel with just sunlight and a catalyst that does not contain heavy metals. Using vanadium, a metal commonly used in steel and aluminum alloys, the team was able to turn polyethylene into formic acid, a naturally occurring preservative and antibacterial agent also used in hydrogen fuel cells and for energy generation by power plants.[33] Of course, you're not about to do that at home. But without the use of high heats, the process may be desirable in some circumstances for labs with potential access to vanadium.

Creative proposals for organic containers have been another option to remedy our plastic crisis, but may never be scalable to a level of societal consideration. Options like banana leaf packaging for goods at the supermarket or seaweed pouches containing sports drinks and alcoholic beverages may be unique and welcome oddities, but simply may not be feasible at large scales.

One compelling new solution has been developed by Goleta, California-based Apeel Sciences, a Gates Foundation-backed company known for creating an edible, plant-derived liquid sprayed onto fruits and vegetables to extend their shelf life without the use of containers (or even refrigeration). The spray-on, plant-based finish reportedly slows down the rate of water loss and oxidation of perishable foods, simultaneously cutting down on food waste and the necessity of air-tight containers. The company was a World Economic Forum Technology Pioneer and one of *TIME* Magazine's 50 Genius Companies. The coating is invisible, does not confer any additional taste, and is FDA-certified as safe. And while the company is tight-lipped about precisely what happens during the film processing, it claims the application is processed simply from organic material. The seeds, skin, and pulp of grapes and other commonly used fruits and vegetables are processed for their oils, which are broken down and mixed into the solution on a molecular level.[34]

Removing plastic from our lives will clearly involve more than just our containers. Plastic-free plant bottles and more sustainable bags at our stores are just a few of the possibilities. But innovations are ongoing. San Diego-based start-up TwentyFifty, named after the year in which plastic waste in our oceans will potentially exceed fish, specializes in producing biodegradable, compostable cutlery consumers can actually deposit in gardens outside. The company's knives, forks, and spoons are made from all-natural grains like corn, soy, and wheat, offering a decomposition time of roughly 30 days' time.[35] Judging by the promo pictures, the cutlery essentially looks like durable, uncooked grain pasta—which is heat-resistant up to 170°F.

Aside from plastic containers and cutlery, what do we intend to do with the rest our garbage? New innovations in waste management center on several great new concepts: the circular economy utilizing more of our waste materials, waste-to-fuel designs, and similar biodegradable prospects from plastics.

Another BEV investment company, Davis, California-based Sierra Energy, uses its FastOx gasification system to incinerate garbage—but does so at temperatures high enough to break down waste at the molecular level and create energy in the process. Everything from household trash to hazardous waste and thrown-away tires, Sierra claims, can be sent to its system and "gasified" at temperatures reaching 2,200°C (4,000°F). Which, for reference, is roughly twice as hot as the middle of an average volcano.

Engineers first inject purified oxygen and steam into the equipment's base. As waste descends through Sierra's gasifier, it reaches progressively hotter temperatures. Different materials break apart and gasify due to the different temperature levels. At the machine's lowest point, any left-over waste is reduced to carbon char, inorganic materials, and metals—most of which can then be shuttled off to different industries for further use. The leftover carbon char, though, is actually used to produce the syngas that helps fuel the process next time around. As the system avoids the nitrogen in ambient air, Sierra can avoid dangerous by-products and greenhouse gases throughout the process. No need for landfills in this case, just modular gasification units able to be placed anywhere feasible.

Even more incredible: one research team at Rice University found that they could essentially "flash fry" any waste carbon source directly into industrial super material graphene. You will recall that graphene is a single, thin layer of graphite—a material used in pencil lead. Graphite is an all-carbon material just like graphene, but the two have different structural arrangements of atoms. Chemists call these kinds of similar yet different materials "allotropes." A diamond is another carbon allotrope because the material is simply a solid form of pure carbon.

Differences in atomic structure will give allotropes different material properties. Graphene's structure confers quite a few beneficial properties: it's 100 times stronger than steel, flexible, impermeable to many liquids and gases—and it is very conductive (heat included, of course). To derive this material from the carbon contents in household garbage would therefore be immensely valuable for making trash more circular.

Carbon-based contents in household garbage will include anything organic, like food waste, as well as most plastics. Through joule heating, the team at Rice exposed waste carbon materials to an extremely high temperature of 3,000 Kelvin (close to 5,000°F). This temperature breaks the carbon bonds within materials, which can then be re-formed into the atomic structure of graphene. Trash components that are not carbon-based are released as gases. The process requires no toxic solvents and can produce high-quality graphene more quickly (and at a lower cost) than many conventional processing methods. The Rice team's "flash graphene" process occurs within 10 milliseconds, quickly turning anything from apple cores to car tires into usable graphene.

Flash frying waste carbon into graphene potentially gives us a viable use for more of our trash. That's great news when the world reportedly throws out anywhere from 15 to 20 percent of the food we grow due to cosmetic imperfections and supply chain mismanagement. Transforming the carbon within food or plastic trash into graphene can also sequester it more effectively than merely allowing it to recirculate after decomposition. The team at Rice was able to produce only a few grams of graphene—which is still very impressive—but not yet at a level of commercial viability for the process. So flash graphene is still in its early stages of development, but we might expect commercial viability as soon as the end of 2021 with continued research.

In another fascinating new "waste-to-fuel" operation, a research team at University College London has developed a method for converting non-recyclable household waste into methanol, which can then be refined to produce an aviation fuel. The team begins with a controlled heating and gasification system, converting the organic components of household waste (or municipal solid waste) into syngas. That syngas is then purified into "syncrude," which is then refined into transportation fuel—all while treating waste in the process. Rather than the plastic to fuel process described previously, this research focuses more on the organic contents in trash.

The research team has estimated that conversion plants could be built near landfills around the UK, potentially treating up to 120,000 tonnes of waste each year while producing a minimum of 22,800 tonnes of fuel. The team further estimated that treating all of the UK region's waste could produce roughly 3.5 million tonnes of jet fuel annually (about a third of the UK's aviation fuel needs)—fuel which reduces greenhouse gas emissions by 70–80 percent compared to their fossil fuel equivalent (equal to taking up to 5.5 million conventional gas-powered cars off the road by 2050).[36]

The concept of harvesting jet fuel from general waste materials is no longer a well-kept secret. Multiple companies now boast these capabilities, and most of them have major partners for expansion. Energy giant BP has already partnered with Fulcrum Energy to produce jet fuel from garbage. Waste is once again gasified first and then introduced to a catalyst that renders the fuel thereafter.

Individual airlines have now partnered with biotech companies capable of these engineering feats. British Airways has partnered with UK-based Velocys to produce jet fuel from rubbish starting in 2024. North American carrier JetBlue has partnered with Finnish (but also Houston-based) waste-to-fuel specialist Neste. In fact, JetBlue was the first American airline to announce they would be going carbon neutral in 2020 on all domestic flights.[37] Admittedly, they will first accomplish this through carbon off-set purchases. But the biofuel JetBlue will increasingly obtain from Neste is reportedly 100 percent produced from waste and residue raw materials, offering a lifecycle carbon footprint up to 80 percent lower than fossil jet fuel.[38] For an industry that will supposedly be tough to decarbonize, that's not a bad start.

Industry giant Honeywell also produces a "green" jet fuel with sustainable feedstocks of algae or camelina (wild flax), reportedly reducing greenhouse gas emissions by 65 to 85 percent compared to fully petroleum-based fuels.[39] Their fuel is produced as a drop-in ready replacement requiring no changes to currently existing aircraft or fuel infrastructure.

While there may only be a few thousand planes flying at a given point in time, they cumulatively produce roughly two percent of the entire world's global carbon emissions.[40] Substituting fossil jet fuel with biofuels and waste-to-fuel blends, and augmenting shorter flights with electrified versions, will be a crucial combination in decarbonizing the sector. We'll also need to ensure there are no "upstream" emissions in producing those materials, should they be shown to exist.

Or how about using wet bio-waste in addition to wastewater for producing biofuels? One international team of scientists reported having converted biowaste like swine manure and food scraps into a biofuel that can be blended with diesel, ultimately sharing diesel's combustion efficiency and emissions profile (meaning the fuel can be compatible with existing diesel fuel infrastructure). Engineering buffs should check the citations for process and method notes, it's a very interesting process (but lengthy to describe).[41]

Innovations abound. One research team has demonstrated an ability to make use of old newspapers in the greener bulk production of carbon nanotubes.[42] It's just one of many ways to do more with less, and discourage useful products that can shave down processing costs from reaching landfills.

There are industries all over the market that stand to become more circular and reduce our overall needs for energy, resources, and carbon. Dutch company Black Bear began because more than 2 billion car tires reach the end of their life with every passing year.[43] That's essentially 2 billion large circular pieces of rubber wasted every year. The company harvests valuable materials inside tires to generate a profit by producing recovered carbon black from waste tires, a crucial component of rubbers in general, inks, paints, and even plastics. Reportedly, the emissions saved from producing these materials through a circular waste-to-value process saves enough CO_2 to match more than 3 million trees.[44] Waste-to-value processes, when successful, help drive a process in which remediation becomes far cheaper than manufacturing whole new materials.

Fortunately, new developments in e-waste recycling may also allow scientists to recover usable rare earth materials from disposed technology. Researchers at the Critical Materials Institute (Ames Laboratory) in the US found an acid-free method for recovering the rare earth materials from shredded hard drives, for example. The team used magnets dissolved in water-based solutions to recover rare earth materials with over 99 percent purity. They believe that industrializing the process could create an entire new sub-field of "urban mining" for precious metals. Clearly, new advancements in technology repair will go much further than repatriating the metals—and good regulations can support more sustainable extraction methods.

Waste from renewable energy tech can be harder to recycle, but it's not impossible. Waste wind turbine blades are often not repurposed, given that glass and carbon fiber composites can be harder to work with. But some of the materials have found homes in playground equipment, among other things, providing at least a few avenues for re-use. One facility in Germany touted as "the world's only industrial-scale factory for reprocessing wind turbine blades," cuts, shreds, and hammers blades into fragments used in cement making.[45] Another exciting prospect will be swapping out as many solar tech parts as we can with biological equivalents, something we'll discuss for more futuristic technology in the final chapter.

HELPING BIODIVERSITY

Many research teams worldwide are focused on saving at-risk species impacted by environmental damage—particularly species impacted by hotter temperatures and more acidic waters. Efforts range from short-term interventions to all-out species relocations. Sometimes, hard-to-relocate species that can't move (like slow-dispersing trees or corals) must be empowered on site. In these cases, scientists tend to look for similar species more tolerant to the changing conditions of their local environments and try to help spread those crucially adaptive genes. Other times, scientists will indeed try to move a species to safer areas. In the direst of cases, they attempt to engineer hardier strains and variants more tolerant to the conditions we seem to be creating.

Some scientists have been noting which trees around the world have more capably handled conditions of heat and drought. Research teams worldwide are collecting samples of what they call "climate survivors" that may have natural genetic advantages to help propagate those species going forward. Such an effort may prove particularly useful if we are considering which trees we want to afforest more than others (while avoiding any kind of afforestation monoculture, of course).

Some species, however, are like our major banks in being "too big to fail" for their respective ecosystems. Bee populations would be one such example. Coral networks are similar in marine environments. Essentially the lowest rungs of ecosystem food chains and biodiversity stabilizers are what scientists are most worried about.

Statistically, more than one third of all coral species worldwide are at a risk of extinction.[46] When temperatures rise and corals become stressed for nutrients to sustain themselves, they "bleach" and lose their vibrant colors. Normally, symbiotic algae live on coral colonies providing nutrients in exchange for a habitat—and the light reflecting from their cellular material creates the myriad of underwater sights we see. Clear corals indicate that their symbiotic helpers are gone.

Coral bleaching events are becoming increasingly common, and coral populations can reportedly recover only every decade or so. While a coral system like the Great Barrier Reef in Australia, for example, has sustained four "mass bleaching events" since 1998, two of them transpired back to back in 2016 and 2017. Research now indicates that if our emissions continue at current levels, local bleaching will occur roughly twice per decade at minimum by the year 2035. By the year 2044, bleaching is predicted to occur annually.[47]

In August of 2019, the future outlook of Australia's Great Barrier Reef was officially downgraded by the country's federal government from "poor" to "very poor." According to Tim Gordon, a scientist studying how the rise of temperatures have damaged corals around the Great Barrier Reef, "they used to be some of the most colorful, vibrant, bustling, noisy ecosystems in the world, but now many of them are eerily quiet, empty gray rubble fields."[48]

Previously unavailable marine real estate that used to be too cold, just adjacent to coral sites, is now opening up as tolerable as oceans warm. So corals are drifting/retreating. Researchers have found that the number of young coral systems on tropical reefs has plummeted by 85 percent, but has only doubled on "subtropical" reefs where the waters are slightly warmer.[49]

Scientists are currently seeking out particularly hardy coral variants exuding tolerance or even thriving in these conditions. They are seeking out super corals, if you will, to help spread in the event that less tolerant variants begin to die out faster. Steve Palumbi in the biology department at Stanford University, for example, has helped further efforts to build resilient new reefs to offset some of those threatened. "Reefs aren't just tourist attractions," Palumbi has pointed out. "They're also biodiversity hotspots that protect coastlines from flooding by absorbing wave energy."[50]

The strategy of breeding heat-resistant corals in labs is referred to as "assisted evolution" or "facilitated adaptation." It's a helping hand for a species that does not know what is happening or have plans to adapt to it. In these efforts, successful colonies of corals bred in labs will later be transitionally transplanted into areas currently beset by higher temperatures.

Once again, scientists first look around nature to see if evolution has provided us a case of advantageously adapted genetics. We can then propagate those genes to help spread them into currently disadvantaged areas. Again, we must still avoid a monoculture of genetic strains.

Bioengineers can also isolate the part of an organism's genetic code that supplies advantageously adaptive characteristics and introduce them into other, similar organisms. Failing that, biologists can attempt to seek out advantageous genetic combinations and potentially scaffold them manually with synthetic biology.

For example, lab breeding of corals that show natural resistance to higher heat is already an option being explored. Scientists are performing tests on different corals that will pre-expose them to milder warming to help improve tolerance in gradients, rather than natural corals being exposed to high spikes in heat. The possibility of using a tool like CRISPR to genetically engineer threat resistance into corals is still, according to researchers, nearly a decade away.[51]

Fortunately, scientists have since discovered 34 species of coral found to be regularly exposed to environmental stressors like low oxygen, low pH, and highly variable temperatures in mangrove lagoons around Australia's Great Barrier Reef. Assessing what makes those corals so extreme should help scientists further their understanding of how best to assist other colonies. Studies investigating Caribbean coral regions have also found species able to survive in acidified conditions greater than those expected throughout this century, further raising the possibility that those advantageous genes can be studied (and potentially even shared).[52] Work like this provides us hope that nature has already battle-tested some coral species able to adapt to warmer and increasingly deoxygenated waters rising in temperature.[53]

Ultimately, the global efforts to save corals are focusing on assessing survivors, cross-breeding corals between different populations, altering coral genes to provide advantageous cover for heat tolerance, and transplanting at-risk coral colonies to new locations.[54]

Australian researchers Peter Harrison and Matthew Dunbabin have recently created an underwater drone designed to disperse especially heat-resistant coral larvae. These corals have not only survived in warmer waters, but tend to flourish there. The research team's larvae delivery drone, "LarvalBot," is already able to disperse about 100,000 tiny specimens on each trip. But the team suspects they will be soon be reaching a plateau of millions. LarvalBot also has been capable of accomplishing additional duties like assessing water quality and conducting "marine pest surveillance" for fledgling coral communities.

Since deploying Larvalbot to seed and nourish baby corals, the researchers have synchronized multiple bots to an inflatable LarvalBoat—to cover more ground. In doing so, the team has moved from a delivery system extending 500 square meters per trial up to about three hectares in just a six-hour span.[55]

One research team working with the Hawaiian government on saving their own corals has focused on removing invasive algae to help budding colonies thrive. Some algae tends to limit reef biodiversity by smothering coral, in the worst cases. The team's strategy has involved using essentially a land-tethered wet-vac, the Super Sucker, to remove invasive algae from coral networks. And for a bonus, divers can siphon the algae away to a waiting barge for delivery to farmers as a supplementary fertilizer.[56]

Corals, like trees, can only drift their colonies via dispersal very slowly. By some estimates, trees can only "move" to northern regions at about one hundred feet per year.[57] In reality, trees disperse seeds and the new seedlings will grow wherever the habitat allows them to do so. Just like corals, only new habitats offering safer temperature ranges will allow for new seedlings to develop. As a result, it almost looks like various tree species are moving away from climate change and migrating north. The problem is that they cannot move quickly enough. Scientists endeavor to save what they can, but some trees will just stay there and wither in suddenly drier climates.

Ultimately, attempting to help species adapt to changing environments is hard to balance. Sometimes our efforts work out; sometimes they blow up in our faces. Scientists have a difficult time gauging how much we should be intervening with the species struggling with climate change. Our record for "species relocations" has therefore been quite hit-or-miss. Sometimes they're incredibly expensive and disastrous. Other times we are essentially playing ecological roulette in dropping species into habitats they may not flourish in. But in all cases, we're desperate to at least do something.

Possibilities still abound. Christopher Preston, author of *The Synthetic Age: Outdesigning Evolution, Resurrecting Species, and Reengineering Our World*, listed several here:

> *Bull trout struggling to adapt to elevated temperatures in high-altitude mountain streams could potentially have a gene for heat tolerance inserted into them. Highly endangered black-footed ferrets suffering from generations of inbreeding could have their genetic diversity increased by the insertion of genes from specimens in museums and frozen repositories . . . Honeybees subject to colony collapse disorder could be genetically enhanced by the addition of genes for the fastidious hygiene traits found in some colonies that have proven successful in keeping hives free from parasites.*

To that last point, scientists at the University of Texas in Austin have also recently reported that engineered bacterial strains can be introduced into bee colonies to vastly reduce the risk of colony collapse.[58] There are many ways to assist the species threatened by environmental changes, aside from outright relocation or intervention.

Realistically, our potential to negatively impact our planet is matched only by our power to remedy it in the positive. Should we intervene gracefully, earnestly, in well-intentioned and well-planned ways, research teams have proved themselves capable of doing great things. Scientists are fairly certain, for example, that genetically engineering American chestnut trees simply to save them from a blight that nearly wiped the species out will be entirely beneficial. Hundreds of scientists have involved themselves in a project spanning 26 years to produce a patent-free blight-resistant American chestnut tree—the first American project to seek approval of a transgenic plant to help save a species.

Simply protecting the natural habitats that support biodiversity can also safeguard species without any bioengineering. New science and new tech may therefore help protect some of the most vulnerable species routinely at risk of displacement due our now yearly wildfires.

The responders who are dispatched to fight the spread of wildfires are legitimate heroes, risking their own safety to bring those fires under control. But we also know that science and technology can significantly mitigate what has already become a multi-seasonal occurrence. With the help of sensor networks gathered into larger and larger arrays, environmental information can help alert responders in advance of impending wildfire conditions. Cues from sensor networks can feed into an artificial intelligence system, comparing their environmental information to weather analyses from the National Weather Service. Advanced systems can compare these datasets with historical trends and produce predictive response options.

One new set of software developed by the WiFire Lab at the San Diego Supercomputer Center corrals data points like these from multiple government organizations to produce real-time, on-the-spot analysis of statistical probabilities for wildfire origins and developments. This data, along with vegetation data from the US Department of Interior; NASA satellite readouts; pertinent topography details that would potentially enable faster spread of fires; flammable materials known to be present in areas logged in various databases; and more can all contribute to stopping wildfires quickly. Powerful new tools like these can potentially help us protect more natural systems and the biodiversity within.

PLANETARY CLEANUP: POLLUTION REMOVAL

If you don't live in a place where smog pollution is a major problem, it can be difficult to imagine how that issue manifests in everyday life. In some Chinese cities, smog is so bad that citizens occasionally receive "red alerts" indicating that schools are closed, flights are cancelled, vehicles should be off the roads, and they should probably stay indoors. In some cities, smog days force the government to issue legal mandates that cars cannot be driven anywhere. Major sports matches are canceled to protect the athletes from the pervasive mists of toxic fog emanating from industrial areas.

Consistent exposure to smog is incredibly unhealthy. One study from the Journal of the American Medical Association even found that routine, long-term exposure to ground-level smog can produce health defects akin to smoking a pack of cigarettes a day.[59] Scientists worldwide are hard at work generating solutions for the air pollution and smog besetting major cities. There are an incredible assortment of new technologies that assess air quality and reduce pollution, but the best solutions seem to be remarkably low tech: plant trees, legislate for appropriate pollution controls in your country's industrial areas . . . and plant more trees.

Other times, research teams and NGOs combine the high-tech and the low-tech in interesting ways. For example, consider that homeowners have long utilized indoor houseplants to clean the air. Now, scientists have helped to improve this process with bioengineered, superior air-filtering varieties.

Scientists at the University of Washington have recently taken genetic material from common houseplants and modified a strain of pothos ivy to include a segment of rabbit DNA. The team theorized that because rabbits are among the few species that can metabolize alcohol like we do, their ability to process airborne, alcohol-based chemicals could be useful for houseplants. Researchers also knew that the protein they'd selected could transform a harmful chemical like benzene into phenol, and chloroform into CO_2 and ionized chloride.

The result? Houseplants that eventually reduced airborne benzene levels in testing conditions by 75 percent in eight days, and reduced chlorine levels by 82 percent in only three days.[60] The chlorine was virtually unde-tectable in only six days. The team further estimated that a hypothetical biofilter created by engineered houseplants would provide clean air at rates comparable to commercial home particulate filters—again, without all the industrial footprint.

One might be surprised to find traces of chloroform in the air at home. But chloroform reportedly enters homes via chlorinated water, while ben-zene dissipates from gasoline and can subsequently mix with the hot air in your shower (or water you've boiled on the stove).

The genius in the University of Washington's team was in dreaming up something of a "green liver" by mixing the genetic information of actual liver proteins with leafy houseplants. The additional foresight was in choosing a houseplant like pothos ivy, tropical and unlikely to flower in the Pacific Northwest lab test site (read: unlikely to spread the modified genetic material via cross-pollination). I suspect we'll see more of these lines of inquiry.

Techniques for addressing air pollution abound. Thailand went hi-tech with aerial EV drones (and even a few small planes) spraying water mist into their lingering plumes of smog to break up polluting particulates throughout 2019.[61] Indian start-up Chakr Innovations partnered with Dell to catch diesel exhaust and filter out soot for black printer ink. The company boasts capturing about 90 percent of the harmful pollutants from smoke-spewing equipment.[62]

Some research teams have taken to attaching air assessment units to the tops of Google Street View cars to exploit the nationwide air data they can obtain.[63] Other teams have noticed that public transit vehicles are almost always navigating city streets—and have attached fine particulate filtering equipment to the front and top of buses. Because why not?

LAST STAND DAMAGE CONTROL: GEOENGINEERING CAMPAIGNS

In terms of bigger efforts to adapt to the realities of climate change, you have probably heard about largescale and consistently controversial geo-engineering schemes. Geoengineering campaigns essentially undertake large, often directly invasive efforts to engineer our planet's climate in ways more advantageous to humans. Or, at least, that's the tagline. Who benefits and how is largely a matter of perspective.

Efforts to combat climate change like mass-afforestation and deployment of direct air capture units are already campaigns to engineer our climate. But "geoengineering" seems to have the further connotation of direct, short-term interventions into climate function.

One example is solar radiation management (SRM), a concept discussed earlier during the section on "cool roofs" that direct more sunlight back out into space. There are actually several SRM strategies being developed around the world at much larger scales, some of which alter the Earth's environment more invasively than others—and more clearly fall under the "geoengineering" category. Pumping aerosols into the air, for example, has been suggested as a low-cost strategy for re-directing more sunlight away from Earth's surface and lowering temperatures quickly.

Geoengineering supporters imagine that if we want a quick, low-cost method of cooling the planet, simply releasing stratospheric aerosols will help lower surface-level temperatures immediately. Others believe that efforts like these will only stifle meaningful progress and potentially cause unforeseen consequences, even if the emerging research on aerosols project low risk.

In fact, a recent IPCC Special Report estimated that releasing aerosol particles into our stratosphere would theoretically offset 1.5°C of warming for only a mere $1 billion to $10 billion per year.[64] Authors writing the IPCC's report noted SRM efforts could help to reduce the risks of sea level rise, sea-ice loss, and even the frequency of extreme storms in the North Atlantic and heatwaves in Europe. On the other hand, they also noted that SRM could lead to changes in regional temperatures for adjacent countries, changing precipitation patterns and potentially reductions in biodiversity. And, of course, they noted that SRM should never take the place of meaningful emissions reductions or carbon capture.

Scientists have been busy studying the ramifications of SRM and other various geoengineering initiatives, ranging from the legal and ethical dimensions to the political and economic. In terms of aerosols, researchers have been busy studying how stratospheric injections might impact local and global climates; whether aerosols pumped into a location like the Arctic might differ from the effects of starting somewhere else; specific types of materials one might use; and more. Governments around the world have studied the potential for initiating immediate, short-term relief from high heats in the form of SRM, including China, India, various levels of government in North America, and throughout Europe.

New assessments have focused on establishing a roadmap for "responsible exploration" of geoengineering to set some ideal ground rules, with stratospheric aerosols potentially being one of the safest. Aerosol supporters believe the effects will likely be similar to what happens when volcanoes shoot ash into the air. One example is the 1991 eruption of Mount Pinatubo in the Philippines that launched a sulfur dioxide mist throughout the stratosphere. The effect, reportedly, was a temporary local cooling of about 1°F over roughly 15 months.[65] That kind of immediate, temporary, local cooling can sound very appealing to people suffering heatwaves, wildfires, or droughts. But, seeding our stratosphere with aerosol particles does little to impact the other effects of excess greenhouse gases (like ocean acidification) and may delay more meaningful action.

Realistically, if low-key initiatives of geoengineering were proposed alongside NETs and alongside vast, rapid decarbonization of our worst emissions sectors, we might have something. But geoengineering enthusiasts are not often making that case.

Going in depth into solar radiation geoengineering is outside the scope of this book, which focuses more on the strategies for reversing climate change—rather than "quick fix" and sometimes sci-fi-based solutions. At this point, the unpredictable nature of modifying the Earth's albedo (reflectivity) in a dramatic fashion through SRM could result in complex and unanticipated effects. Researchers are still a great many years from ruling definitively on the full repercussions.

Changing the "radiative balance" in varying levels of our atmosphere and micromanaging heat reaching and leaving us can create a number of unpredictable variables. Phenomena like wind patterns, temperature ranges between different areas, evaporation of water from the ocean, behaviors of plant species in response to sunlight reaching them . . . all of these factors and more can create interlocking and unpredictable outcomes.

There are other geoengineering schemes from mild to wild—from placing parasols above our planet to placing barrier walls against the Arctic sea ice to prevent warmer waters from reaching the ice itself (and preventing its melt in the process). Using a deeper sea pump to bring cooler waters to the surface near larger Arctic ice shelves has been another idea. Or seeding clouds with sea salt to make them more reflective against sun-

light. Or manipulating the vapor density of clouds with lasers. These are
the kinds of plans you might imagine super-villains hatching in a secret
lair somewhere, rather than the collective efforts of academic teams
worldwide.

Some of these options—if undertaken—have been criticized for being
implemented too quickly, with too little study and without a crucial level
of public support. Among the more controversial geoengineering options
for ocean carbon uptake, for example, was to stimulate an artificial plankton boom of growth several years ago. Essentially, this option works by
spreading powdered forms of pro-plankton nutrient elements like iron,
phosphorous, or potassium. Phytoplankton near the water's surface then
proliferate and, in the process, absorb greater amounts of CO_2 during
their growth cycle. As a purported bonus, the increase in phytoplankton
is essentially feeding the lowest rung on the food chain, which should lead
to increased opportunities for organisms higher up on the chain. The
proposed result of this approach is that a portion of the carbon these tiny
organisms have absorbed should sink to the bottom of the ocean with
them when they either die or are consumed. This trickle-down effect, it
is hoped, will result in more net carbon being absorbed than is spent fueling the process.

In 2012, a controversial project off the west coast of Canada dumped
about 100 tonnes of iron sulphate into the Pacific Ocean to accomplish
precisely this. Analysts specializing in international law believe that such
a project actually contravened, among others, UN laws prohibiting dumping wastes at sea. Scientists are still not even sure if the effort produced
the intended results. There have been about a dozen other such experi-
ments worldwide to date, but the results are still being assessed.[66]

All of this tells us that vast geoengineering initiatives are still quite a
few years away from serious implementation, if they are to be deployed
at all.

CHAPTER 11

POSSIBLE FUTURES

■

The Pliocene called. It wants its CO_2 levels back.

—RUSSELL MCLENDON, MOTHER NATURE NETWORK (2019)

The year 2050 is often envisioned as the year we'll have finally gotten our act together. It is, among most of the academic literature this author has been able to review, a significant calendar benchmark—the year by which we'll have finally spent decades creating global, systemic action to address climate change. Brighter scenarios often see us far beyond the race to zero emissions, way beyond the denialist debates, beyond all the chronic political short-termism, and certainly beyond the fractured, isolationist policies of countries currently in it largely for themselves.

The more optimistic among us believe this century's twenties will probably be even more roaring than the last, and that incredible, international grit and determination are just around the corner. We see the seemingly incredible ascent up the "Mount Impossible" of zero emissions broken up into a much more incremental march up the Mount Improbable of single-digit reductions—year after year after year.

The roadmap toward 2030 is a little more certain. For example, it will likely be filled with continuing cost reductions for renewable energy sources. Tech advancements across the board discussed throughout this book have made the scenario more likely. There should, throughout the 2020s, be much-needed injections of research funding for everything from geothermal renewables to clean but non-renewable nuclear fusion and green hydrogen (to say nothing of biohydrogen). Large networks of tidal energy arrays will outpace even offshore wind farm installations. Cities will increasingly integrate their energy portfolios with all of the above, along with distributed solar tech, kinetic footfall pads and IoT automation managing energy efficiently. Green spaces and "pocket parks" will likely be built into expanding urban divisions with large, vertical forests lapping up sunlight and CO_2 for the increasing number of urbanites worldwide.

Government-sponsored training programs for those still working in fossil fuel dominant industries will likely become the obvious solution for obstructionist political voting. Responsible policy decisions like 45Q should give life to new industries and widespread carbon capture. Expanding construction projects for new electric-charging infrastructure and grid upgrades will continue all around multiple countries. Retrofitting programs for outdated buildings will reward owners with net-zero status (and energy/resource cost savings), invigorating regenerative architecture and locking away mass amounts of carbon as we outfit smarter cities with smarter features. The legally mandated by-law requirements for zero emissions buildings will appear in more cities worldwide.

Renewable energy inputs should be standard on all new homes—but not because of some eco-liberal agenda. The ability to not only pay for one's home energy bills but also to bank and even sell excess energy is simply too seductive for homeowners. Whether we're talking about advanced new solar shingles, thermal energy tech, or a combo of solar and radiative cooling tech; it doesn't really matter. New homes will have to feature one or some of these options to remain competitive with others. Legal mandates for new homes outfitted with energy tech will likely expand from California to practically everywhere.[2] Plummeting prices of renewables should make this a non-issue.

By 2050, we will have long since solved the "intermittency" problem of current renewables with widespread, interconnected, distributed energy sources feeding into smart grids—and have ready access to backup power through river-based hydro, tidal inputs, geothermal energy, pumped hydro, compressed air, molten salts, other P2X equipment like gravity battery networks, and advanced lithium ion storage solutions. Smarter grids will connect greater portions of countries, leading to more sharing of renewable energy in times of excess among neighboring communities. International grids will provide even more chances to share during times of excess. Smarter grids will also allocate energy more efficiently and allow operators unparalleled energy management in real time.

Direct air capture should ideally, at that point, be a subsidized public service alongside a robust industry to speed up our climate response. Carbon dioxide is ideally "upcyclable" enough to be captured in meaningful amounts—while publicly funded facilities help to pull down atmospheric CO_2 for storage to help us meet our climate goals. The subsidy is seen as a humanitarian effort, where richer countries offer meaningful global aid (particularly as climate damage costs have been calculated to far outweigh the subsidy costs of helping to draw down atmospheric CO_2 more quickly). These efforts are clearly happening amidst the rapid expansion of clean energy sources to cut emissions as quickly as possible.

We'll also feel a lot of pain along the way. The years along our road to zero emissions will undoubtedly feature more extreme weather events, fires, floods, drying aquifers, and so much more. Superstorms we haven't named yet will batter coastal communities and low-lying cities. Those who used to try to rebuild following floods will have given up long ago, after annual floods turned into annual toxic submersion. Governments will have long stopped bailing out homes and businesses in these locations, and most banks no longer insure them. Potentially thousands of species won't live to see it all, despite our growing efforts to save them.

By 2050, cultured meat should no longer be any weirder than almond milk once was. The simple, sensory simulacra of enjoying something that tastes good and provides the same health benefits of traditional livestock seems like a no-brainer. "Real leather" products should be about as necessary as exotic cigars are for smokers—sold for prices likely only connoisseurs will find palatable. Food production should be hyper-local far before 2050. The average amount of transport for the American meal from field to plate should no longer be approximately 1,500 miles.[1]

Synthetic biology will expand its realm of expertise in coalition with molecular engineering to tailor advanced new designer organisms. These organisms will help produce drop-in fuels for millions of new motorists worldwide, and simple custom-made products on demand. All of the plastics produced, at this point, should be bioplastic. Batteries may literally be alive, as may parts of your new vehicle.

Our capability for designing sophisticated new microbial life will likely have long ago extended to designing *multicellular life*. We'll start with what is essentially unfolding cellular origami and AI-assisted computational biology will help take it from there. Indeed, this has already begun in earnest.[3]

There are, however, some important things to consider. The hunt for sexy technological remedies for our problems cannot be a default. Meddling in the affairs of other species and ecosystems cannot be our assumed privilege. Designing and utilizing life for our own self-sustenance should come with a good measure of gravity. Because while it's nice to believe we'll always do great and responsible things, there are times when we certainly do not. We must have the humility to check that balance.

Right now, in the year 2020, there's plenty to be excited about. As you read these words, nearly three dozen cities worldwide have already reached *peak emissions*, and their total CO_2 emissions are in decline. An analysis from the C40, a coalition of 94 of the world's largest cities, confirmed that 30 cities have already reached this goal. Cities include Athens, Austin, Barcelona, Berlin, Boston, Chicago, Copenhagen, Heidelberg, Lisbon, London, Los Angeles, Madrid, Melbourne, Milan, Montréal, New Orleans, New York City, Oslo, Paris, Philadelphia, Portland, Rome, San Francisco, Stockholm, Sydney, Toronto, Vancouver, Venice, Warsaw, and Washington, DC.[4] Moreover, the analysis revealed that since reaching peak emissions levels, the drop has continued: the 30 cities in question continued greenhouse gas emissions reductions by about 22 percent on average.

Crucially, research already tells us that positive energy and climate policies can actually do what they intend. Research published in *Nature Climate Change* has shown that among the countries where emissions have decreased the most between 2005 and 2015, the countries with the most prominent emissions reductions were those with the largest number of energy and climate policies in place.[5] Lobbying your political leaders to make climate-conscious decisions clearly can pay off.

It is great news that cities seem to be particularly adept at making large-scale emissions reductions on smaller time scales. Consider that the UN predicts nearly 70 percent of humanity will live in a city by the year 2050.[6] To achieve peak emissions sooner rather than later, the actions of dedicated cities may be able to circumvent at least some of the lag time created by federal policy-makers. Further, the expansion of cities will give us a chance to really make the most of new carbon-locking products like cement and mass timber. Researchers have estimated, for example, that urban areas will increase by up to 1.3 million square kilometers (over

500,000 square miles) between 2015 and 2050, given our current pace of urban expansion.[7]

In early 2020, the IEA actually reported that global emissions from energy production had already flatlined. Which is to say, they did not grow.[8] That's great news, of course—but this figure stands to grow significantly again if more developing communities come online with fossil fuel use. The IEA further reported that moving toward renewables will also confer several key benefits in our fight against climate change. According to the organization's Renewable Energy Roadmap (REmap), accelerating our energy transition will yield immense capital savings for global governments in not paying for climate damage recoveries, or fossil fuel energy subsidies and the health costs associated with our current trajectory. By their own estimates, the global economy will save roughly $160 trillion USD simply not subsidizing Big Oil or paying the hospital bills of people suffering from environmental damages.[9]

It's true that this book has cited reports claiming that things may be getting worse faster than expected. However, some things could also get better faster than expected. Previous studies of solar capacity worldwide have not often factored for new floating pavilions of solar pads or tandem perovskite and silicon cells increasing consumer demand. After all, who knew? Most calculations about the applicability of biofuels have failed to factor in engineered strains of cyanobacteria capable of making drop-in fuel stocks. Arguments about whether using croplands for biofuels will stunt the agricultural sector never seem to factor in vertical farming or bioengineered supercrops. Not because these things don't exist, or that they're just moonshots. Rather, they have simply been very difficult to predict. Estimates about the rate and efficacy of afforestation efforts do not typically factor in drone swarms planting them by the tens of thousands per day.

My point is not to sell promissory science. In fact, research tells us that faith in technology can be dangerously debilitating when it delays meaningful change. One team from the Lancaster Environment Centre recently analyzed decades of political action (or inaction, more accurately) linked to attitudes of what our climate goals should be. They found that technological promises have consistently yielded moving goalposts for what we intend to accomplish with climate action. The team found that, "for forty years, climate action has been delayed by technological promises . . . Each novel promise not only competes with existing ideas, but also downplays any sense of urgency, enabling the repeated deferral of political deadlines for climate action and undermining societal commitment to meaningful responses."[10]

Let's be clear: meaningful social and cultural changes must also take place to live in a more sustainable world—full stop. Technological remedies cannot allow us to slow ourselves down in ushering important societal changes like stopping deforestation, increasing the ratio of plant-based foods in our diets, enabling mass-electrification of transportation (which we have technology to do right now), and more.

However, in terms of empowering social changes—like eating less conventional meat products and transitioning more toward plant-based and cultured meat alternatives—technology will help us to do what has thus far proven exceedingly difficult. Many of us carnivores simply love the experience, making any kind of systemic transition seem out of reach in the near-term unless policymakers step in. New developments in scalable alternatives will help here significantly.

Every ounce of conventional meat that we can replace with a plant-based or cultured alternative will help free up that land for important projects. Independent think tank RethinkX has estimated in their 2019 Sector Disruption Report on food and agriculture that by 2035, 60 percent of the land currently used for livestock and feed production will be freed for other uses. That amount of land tallies to about 485 million acres. That's 13 times the size of Iowa. If all of that suddenly free land was converted for reforestation with techniques maximizing carbon soil storage, researchers estimated that "all current sources of U.S. greenhouse gas emissions could be fully offset by 2035."[11]

Naturally, that tally is probably a bit optimistic. But when we imagine that land freed from livestock use can potentially contribute to carbon storage through reforestation—in more and more countries worldwide, thanks to meat alternatives made possible by new technology—our emissions reductions outlook suddenly looks slightly brighter.

My point is ultimately to dissuade doomsaying in the face of mounting evidence that transformative new science can help us significantly. It is realistic hope. We are designing tailor-made proteins with the help of artificial intelligence. We're designing new enzymes. The world's first all-synthetic cell was designed already—ten years before the writing of this book.[12] With these developments operating in mutually reinforcing ways, new and previously unimaginable biofuels and bio-based manufacturing seem more probable than promissory on a timescale of 2050.

There will undoubtedly be incredible new scientific developments between the editing and release of *Our Livable World*—in anything from nanoscience to molecular engineering—that could revolutionize the way we see energy. And they could come right out of left field. Scientists have already bioengineered microbes to create a myriad of useful chemicals simply by living in vats of water—and have already created fuel for cars from nothing more than sunlight and air. But newer advancements could become downright mind-blowing in years to come, given the latest projects emerging from R&D labs that we haven't even discussed yet.

Consider that scientists are currently working on technology that will integrate "electroactive" microbes into batteries, fuel cells, and solar panels. Doing so will create a medium for producing energy with far fewer rare earth elements and less toxic equipment—and will generate and store energy more cleanly than ever before.

Ever since 1988, scientists have known about so-called "exoelectrogenic" (electricity-producing) bacteria like *Shewanella oneidensis*. The Shewanella microbe was found to breathe by transferring an electric charge to metal materials outside of its cell membrane, "respiring" electrons to shed them. That electron cycle was simply part of the organism's normal metabolism, but implied to enterprising scientists that there may be excess electrons there to harvest for use. And over the decades, researchers have been exploring ways to generate sustainable energy with the help of microbes alone. Only recently, with the tandem developments of next-gen nanotechnology and bioengineering, have new possibilities opened up to realistically accomplish this goal.

We can think of our metabolism as powered primarily by an electron transport chain. Animals like us consume glucose, which supplies electrons—and the oxygen that we breathe receives electrons, creating a flow for basic metabolism. Animals that breathe oxygen to survive and use aerobic metabolism require an electron transfer as part of a trade-off with inhaled oxygen and water production. What is different in a microbe like *Shewanella oneidensis* is shuttling electrons in the absence of oxygen to an adjacent receptor material outside of the cell (which acts as the final electron acceptor), transferring a charge in the process.

To harvest this electron flow, scientists have attempted to fasten electrodes to a scaffolding upon which the bacteria might live. Thus far, materials used to shuttle electrons have been too inefficient, not as conductive as necessary, or too inhospitable for the bacteria to thrive on. More recently, however, scientists have developed several different materials and concept designs to both shield Shewanella and create a flow of electrons sufficient to power equipment. In other words, creating a "biohybrid" device housing bacterial energy generators supplying power largely for free. The result, if scientists succeed and scale up a viable device, will be another new, emissions-free energy source.

One approach has been to pair Shewanella with a nanocomposite material that supports exoelectrogen growth while simultaneously channeling electricity in a controlled manner. A team of scientists from the Karlsruhe Institute of Technology (KIT) in Germany recently created a porous hydrogel in which exoelectrogenic bacteria can be placed, equipped with carbon nanotubes and silica nanoparticles to conduct electricity. A nutrient-filled medium is added to the mix to keep the bacteria alive while they produce the electricity shunted to attached electrodes.

Scientists ultimately found the bacteria successfully grew, attached to the material, and filled the pores of the hydrogel along with the nanocomposites and silica nanoparticles.[13] The power output was relatively underwhelming, to start. But scientists believe that with some genetic scaffolding and the application of synthetic biology, electroactive microbes can potentially become biological electric generators able to pump out energy for the price of culture nutrients.

Another approach has been to create what are essentially *biological solar panels*. So-called "biophotovoltaics" (BPV) pair *Shewanella oneidensis* with a microbial partner producing the nutrients they need—and when their partner lives on sunlight alone, the result is a biologically driven solar power device.

In one recent BPV design, photosynthetic cyanobacteria groups were engineered to produce higher yields of a chemical called D-lactate that Shewanella then uses to live. Sunlight provides energy for the cyanobacteria, which provides energy for Shewanella, powering electronic equipment thereafter.

BPV energy systems hold immense promise as an emerging technology. Because while the power output for BPV applications has been lower than traditional solar panels to date, the prospect of bioengineering a greater output is possible. Indeed, scientists from the Institute of Microbiology of the Chinese Academy of Sciences recently reported that their biopho-tovoltaic system had broken new records in operation time and power output.[14] Moreover, unlike traditional solar panels, BPV systems do not require toxic and difficult-to-recycle semiconductor materials strip-mined from the Earth. The cyanobacteria involved in BPV systems also will be using CO_2 to help produce electricity.

It will be crucial to off-set at least some of our impending reliance upon traditional solar panels. Solar PV tech is expected to increase dra-matically in years to come as more nations have the means (and the de-sire) to install them. But this will also mean an increase in the rare earth elements used in their manufacturing. These materials—also crucial for renewable energy techs like EV batteries and wind turbines—currently require an environmentally damaging extraction process to acquire. Ex-traction also sometimes involves working conditions with gross human rights violations in countries like the Democratic Republic of the Congo, where children as young as six are sometimes sent to toil in the mines. That must all change, if the world is to scale up more use of conventional solar cells, wind turbines, and EV batteries. Suitable replacements (or at least ethically sourced materials) must overtake the current components of conventional solar tech—and a tandem of cyanobacteria and electro-active bacteria may help to achieve part of that.

New innovations will likely increase over the years with the help of AI. Machine learning can potentially speed up the discovery of new materi-als, for example, by sifting through the countless chemical structures in our records for any of the properties we desire. Academic teams used to achieve that goal through laborious planning and trial-and-error experi-mentation. Now, scientists have already taught an AI to understand the fundamental laws of quantum mechanics.[15] That AI then proved capable of predicting the molecular wave functions and electronic properties of molecules which, in the words of the research team involved, could "dras-tically speed up simulation programs for drug molecules and new materials."

In early 2020, an AI-powered super-computer helped scientists at the University of Vermont and Tufts University create the world's first living, self-healing robot assembled from stem cells.[16] Although the term "robot" doesn't perfectly apply: as a bundle of variously formed cell types, the bot is neither a traditional robot nor a species of animal. Scientists simply called it a xenobot—a living, programmable organism.

Scientists used advanced AI to cut and reshape stem cells into specific "body forms" never seen in nature. With the help of computational biology, scientists were able to produce various cell combos (skin and heart muscle cells) resulting in unique behaviors for their bots. Designs yielding successful functions were then banked, while others were discarded. Scientists were able to essentially program predictable behaviors into their new creations. According to the xenobots research team, their bots could be used "to clean up radioactive waste, collect microplastics in the oceans, carry medicine inside human bodies, or even travel into our arteries to scrape out plaque."

Future technologies are already crossing the spectrum from evolutionary to revolutionary. It therefore seems unimaginable that we cannot find our way out of climate change. There really is, as Dr. Kate Marvel has noted, a continuum of futures and a continuum of possibilities. Or as best-selling climate journalist David Wallace-Wells wrote in his sobering work *The Uninhabitable Earth*: "The emergent portrait of suffering is, I hope, horrifying. It is also, entirely, elective."

Historically, it took us about ten years to get to the moon after we decided to go. Our current smartphones now have much more computational power than all of NASA had to make that happen. Entire new fields of science needed to be cultivated and explored to solve the many puzzles and challenges of jet propulsion and space flight. We're there, right now, with a zero emissions goal to reach.

Great things can happen in ten years. In 2011, when the US Department of Energy launched its SunShot Initiative to reduce the cost of solar power by 75 percent before 2020, it hit the target three years early.[17] This is why some have called for a similar CarbonShot Initiative with the same kinds of incremental goals in mind.[18]

When we discovered there was a massive hole in our planet's ozone layer, and that failing to address it could be disastrous for life on Earth, nations came together, rolled up their sleeves, and got to work. The Montreal Protocol, signed into effect in 1987, prohibited the production of chlorofluorocarbons (CFCs) in refrigerant materials. Phasing these materials out of production was a solid step toward fixing the problem. In fact, one satellite analysis released in early 2020 found that the ozone hole was well on its way to healing. It is expected to fully recover during the 2030s in the Northern Hemisphere, and by the 2050s for the Southern Hemisphere.[19]

New research shows that addressing the hole in our ozone might have actually been our first global effort to directly lower the Earth's temperature. According to scientists, because CFCs are thousands of times more potent a greenhouse gas than CO_2, the Montreal Protocol not only saved the ozone layer, but also kept the planet an average of 1°C (1.8°F) cooler—up to 4°C (7.2°F) in the Arctic—than it would otherwise have been.[20] Ironically, scientists mused, the Montreal Protocol was even more effective at addressing global warming than agreements designed specifically to curb greenhouse gases.

What form could new international efforts take? One easy prediction may be the increasing prominence of cross-national energy grid systems. These systems already exist, to an extent. Just as larger countries share their regional grids between states or provinces, so, too, will we likely see these applications deployed in cross-national grid applications. Sharing renewable energy inputs across larger distances will undoubtedly help to smooth out the intermittency issues of relying on them for base power. Because even if the wind isn't blowing in Texas, for example, it might be blowing in Arizona. And even if the sun isn't shining on the East Coast, it might still be shining on a West Coast enjoying more solar energy than they require.

One study published in *Nature* found that transitioning from a regionally divided electricity sector to a national system enabled by high-voltage, direct-current transmission could slash CO_2 emissions from the electricity sector up to a stunning 80 percent relative to 1990 levels.[21] The study's authors declared the reductions are possible "with current technologies and without electrical storage." In fact, climate scientist Katharine Hayhoe recently reiterated the calculation that if the US were to group together clean energy installations as a land mass, the country would only need about a square of roughly 100 to 120 miles per side. That is comparable to the land area currently used to supply maple syrup or play golf.[22]

Or consider a 2015 headline from the *Guardian*: WIND POWER GENERATES 140 PERCENT OF DENMARK'S ELECTRICITY DEMAND.[23] Headlines like these can sometimes give us a false sense of hope regarding the power of renewable energy for 24/7 coverage, but they can also highlight the promise of international grids sharing peak energy hours. Denmark's grid, in particular, is part of a much larger Nordic Synchronized Area including energy inputs from Germany, Norway, and Sweden. Hydro power coming from Sweden and Norway helps to supplement the others while Denmark's prominent wind capabilities off-set. Energy storage is provided in part by water pumped uphill and then released to flow downhill onto a turbine system in times of need.

Systems like these help to rationalize the sometimes-maligned efforts in a country like Germany transitioning so heavily into renewables. Leaders may well now be looking to tap into expanding energy export options as new transmission lines come online.

Energy expert Daniel Kammen pointed me toward the new (and particularly aggressive) Chinese Belt and Road Initiative as one example of how cross-national energy systems could play out. A play on the Silk Road connecting Ancient China to the rest of the developing world, Belt and Road will be an effort to accelerate and enhance trade infrastructure along Asian countries linked by transport and digital connectivity. China has invested a staggering half-trillion dollars at present to develop more infrastructure along the route, sometimes employing what critics view as predatory lending to vulnerable countries hoping to join in the growth. We should not be surprised if energy-sharing plays a role in developing the economics.

Local energy transmission projects already in the works imply this will be a reality. The Changji-Guquan transmission line in China is years into development and will be capable of transmitting 12 GW of power—meeting the needs of some 26.5 million people—extending nearly border to border at 3,293 kilometers (2,046 miles) long.[24] The energy transmission is roughly the energy needs of the country of Romania (or half of Spain in its entirety). The distance is comparable to traveling from Los Angeles to Cleveland. This colossal project was made possible by the mounting studies implying such a system would yield positive outcomes. Studies investigating the applications of a potential "supergrid" connecting China, Japan, South Korea, Mongolia, and Eastern Russia had already demonstrated cost-effectiveness.[25]

What might a larger-scale international cross-border supergrid look like? Engineers have already speculated about (and subsequently stepped back from) massive global energy connectivity projects in the past. One project in particular, Desertec, was first proposed back in 2009 and planned to connect the sun-rich deserts of North Africa to eager energy customers throughout Europe.

As it turns out, a massive solar installation covering just a portion of the Sahara Desert would reportedly provide 7,000 times the electricity needs of Europe.[26] Blanketing a much smaller portion of the desert could therefore provide Africa all the energy it needs, as well as providing "solar exports." Coupled with connections to wind turbine farms and concentrated solar installations, researchers even speculated that a project of that scale would influence regional climate conditions. Precipitation would more than double. Vegetation would potentially increase by 20 percent.

Unfortunately, the Desertec project ultimately suffered from numerous engineering pitfalls. Ever-present airborne sand was certainly one of them, along with the issue of transporting electricity well over a thousand miles away. China's current long-distance electric cables connecting faraway provinces come close, but the cumulative power of multiple risks ultimately stalled the effort. While the Desertec project was subsequently abandoned, lower prices in renewable energy tech have revived similar efforts.

Nur Energie's TuNur solar power project in Tunisia similarly aims to take advantage of the region's excess of solar radiation to export energy up to Europe for a net gain. Prospects are slightly more realistic, in this case—Nur Energie will benefit from the completion of the Malta-Sicily Interconnector, a 120-kilometer (74.5 mile) underground and underwater power connecting Malta to Europe's grid. Time will tell if Nur Energie is successful, but I suspect the consistent decline in cost for renewables and increasing efficiency of the technology will make such a project inevitable in some form.

Of course, while the African continent is ripe with energy exports, there is certainly a conversation to be had regarding who will get all the energy produced there, and what locals have to say. Some sociology scholars have called projects like these "climate colonialism," not only because of the lack of justice for countries most affected by climate change, but the potential for oppressive power grabs of energy resources by stronger countries.

By some analyses, largescale solar energy exports could be a better fit for the Arabian Peninsula. The area already has all of its trade partners lined up, is already used to exporting plenty of energy, and could utilize pre-existing ports and pipes to ship out renewably produced hydrogen in the future.

One report has proposed utilizing parts of the Maghreb region in North

Africa—including parts of Morocco, Tunisia, Libya, Mauritania, and Algeria—for largescale direct air capture deployment.[27] The team found that using just 0.08 percent of the region could account for removing 10 percent of the planet's annual CO_2 removal, while also generating some water from the surrounding air.[28] This effort would reportedly constitute about 0.27–0.50 percent of the global GDP at mid-century. And in the words of the study's authors, it would be "well affordable for the global society if a climate change survival strategy is the clear target of the people of the world." Again, this would be dependent upon the needs of those who live in and around those areas. There are some indications that Africans may not want parts of their homeland to be an out-sourced climate remedy.[29]

Western Europe and the Americas may one day benefit from their own largescale energy megaprojects. For example, research has revealed that if engineers were to establish wind farms further offshore from each coast in the North Atlantic Ocean, turbines could benefit from stronger wind speeds, producing power exceeding land turbines by a factor of three or more. It may therefore eventually become a compelling option to establish large wind farms in this area with underwater cables bringing the energy home. Researchers have estimated that if humanity were to build a wind farm roughly the size of Greenland in the North Atlantic, the energy produced could realistically power "the current global primary energy demand" during good seasonal months—and at least the energy needs of the entire European Union during seasonal low times.[30]

Of course, scientists aren't about to go building a project the size of Greenland. However, the logistics of scaling wind farm manufacturing and deployment to this level, in the middle of the ocean, is not really what the researchers envisioned. This data simply makes a case for expanding more offshore wind coverage to supply what is apparently far more efficient and predictable wind patterns generating greater amounts of energy. It is not hard to imagine that a wind farm project off the North and South American coasts may supply energy to multiple grids.

Again, these are not quick fixes—just more silver buckshot. The point is to begin the conversations. Keep the research going. Trial test new efforts. Start small when proposals are speculative. Reduce emissions above all. The point, simply, is to get the ball rolling so we don't let future generations down. This is our moment and we still have the time to solve the climate crisis.

In a speech at TEDxStockholm, climate activist Greta Thunberg gave a stirring account of how not only her generation, but future generations will wonder in bewilderment about why today's generations in maturity did not do more to remedy the climate crisis. Her address ended on a somber, pragmatic note:

> *Now we're almost at the end of my talk, and this is where people usually start talking about hope, solar panels, wind power, circular economy, and so on, but I'm not going to do that. We've had thirty years of pep-talking and selling positive ideas. And I'm sorry, but it doesn't work. Because if it would have, the emissions would have gone down by now. They haven't.*

As usual, Greta's words have a knack for piercing through the illusion that just because you're older, you know better. But pep talks about solar panels and wind turbines have not always benefitted from powerful data regarding just how much techs like those truly can solve a lot of our problem—nor a half-dozen new storage and backup solutions to fill in the gaps. Pep talks about a circular economy have not always had the benefit of today's bioengineering capabilities or the results from labs turning our garbage into jet fuel and our car tires into graphene. In short, pep talks have sometimes sounded more like illusory hope when they need not with more supporting data in hand.

Being a techno-optimist does not necessarily mean being a techno-utopian. Plenty of our ideas over time have also been poor ones and it's entirely appropriate to admit that. As noted, we once blew a hole in our planet's ozone layer just by packing freon into our refrigerators. Advancements in fertilizers have allowed us to feed a planet full of people, but they have also given us guns and bombs and worsened our changing climate. The internet has led to an unfathomable level of information sharing between cultures, but it is also being used to produce a post-truth society in parts of the world.

To achieve the seemingly impossible, we must believe that our species is capable of doing good things with the great ideas we have. I believe the pep talk of positive ideas can help significantly—that is, when they are backed up by well-supported scientific research. Positive ideas can become powerful motivators for enacting the changes we believe will help the most. They help to provide us a clear frame of mind for supporting organizations and political representatives who genuinely champion those ideas. What pep talks couched in real research provide for this author, personally, is a sense of hope about our prospects that does not seem anchored in illusion.

It is my hope that the incredible news of evolving sciences and emerging technologies throughout this book has helped to do the same for you.

ACKNOWLEDGMENTS

This book was slowly completed over the course of five years. In the beginning, my work began as a pet project of compiling the many new tech developments offering me legitimate hope that we may one day "solve" the climate crisis. Year after year, I found myself gathering all the latest research that gave me a brief reprieve from the doom and gloom of mainstream media coverage. The effort was more of a self-administered remedy—an antidote to the narrative that we were simply too far gone and too apathetically uninterested in saving ourselves from a world increasingly on fire.

I soon found myself unable to keep all the positive new research to myself. And eventually, in early 2019, a good friend of mine implored me to share some of the medicine. The phrase "I didn't know they were doing *that*!" was essentially our new catchphrase. Every few days, it seemed that breaking research was inching us closer and closer to real, practical hope regarding green tech and climate action. She would later convince me that perhaps other readers out there may enjoy hearing about all these new developments as well.

My first acknowledgment should therefore go to my friend April for her encouragement in compiling this research into book form. But thanks also—obviously—to the many scientists out there toiling thanklessly to save our species from the worst parts of ourselves. Science stands, as Carl Sagan once wrote, as a candle in the dark—illuminating our path forward. Every scientist spending their Monday to Friday grinding away under fluorescent lab lights is among the heroes who will eventually get us out of this mess we've made. More specific thanks goes to academic giants like Howard Herzog, Daniel Kammen, and Michael Oppenheimer for taking the time to speak with me and for keeping the project more grounded in realism (even if they may cringe at some of the more speculative, sci-fi tech parts). I have attempted to avoid undue sensationalism and hope their commitment to strong empirical evidence, wherever it may take us, was present in this work.

I'd also like to thank my first agent, Deborah Hofmann, for taking on this project and helping me wade into the scary world of international publishing houses. Her unwavering enthusiasm for the book was *the* reason that several publishers were interested in following up on it. Although on that note, I also must acknowledge the Editor in Chief at Diversion Books, Keith Wallman. Keith has been incredible to work with and his passion for the power of human ingenuity is one I share. I am therefore very lucky to have worked with Keith specifically—his guidance has been invaluable and he has helped to keep this work from ever feeling like work.

SOURCE NOTES

PREFACE: COVID-19 AND THE ENVIRONMENT

1. See an assessment of the research here: https://www.carbonbrief.org/analysis-coronavirus-has-temporarily-reduced-chinas-co2-emissions-by-a-quarter.
2. See commentary from several experts on this phenomenon here: cnn.com/2020/03/16/asia/china-pollution-coronavirus-hnk-intl/index.html.
3. https://theicct.org/sites/default/files/publications/ICCT_CO2-commercl-aviation-2018_20190918.pdf.
4. https://blogs.ei.columbia.edu/2020/04/28/measuring-new-york-city-emissions/.
5. See researchers at Columbia University discuss this finding and others with the BBC: https://www.bbc.com/news/science-environment-51944780.
6. https://sciencemag.org/news/2020/02/completely-new-culture-doing-research-coronavirus-outbreak-changes-how-scientists.

CHAPTER 1: SCENES FROM THE CLIMATE CRISIS

1. https://docs.wixstatic.com/ugd/148cb0_a1406e0143ac4c469196d3003bc1e687.pdf.
2. https://time.com/person-of-the-year-2019-greta-thunberg/.
3. https://www.cnn.com/2020/01/04/australia/australia-red-sky-fires-intl-hnk-scli/index.html.
4. Quoted in: https://www.yaleclimateconnections.org/2019/12/the-high-and-low-points-for-climate-change-in-2019/.
5. ipcc.ch/site/assets/uploads/sites/2/2019/02/SR15_Chapter2_Low_Res.pdf.
6. http://www.globalcarbonproject.org/carbonbudget.
7. http://www3.weforum.org/docs/WEF_Fostering_Effective_Energy_Transition_2019.pdf.
8. https://www.carbonbrief.org/analysis-uk-carbon-emissions-in-2017-fell-to-levels-last-seen-in-1890.
9. https://www.bloomberg.com/news/articles/2019-08-27/solar-wind-provide-cheapest-power-for-two-thirds-of-globe-map.
10. https://www.weforum.org/agenda/2018/03/chart-of-the-day-the-world-will-add-70-000-solar-panels-every-hour-in-the-next-5-years/.
11. https://www.sciencealert.com/it-s-official-atmospheric-co2-just-exceeded-415-ppm-for-first-time-in-human-history.

CHAPTER 2: LET'S FACE IT: OUR PLANET IS CHANGING

1. Although there are discussions about this figure and why even small discrepancies can matter: https://www.forbes.com/sites/uhenergy/2018/09/07/exactly-how-much-has-the-earth-warmed-and-does-it-matter/.
2. https://library.wmo.int/doc_num.php?explnum_id=10108.
3. For a handy graphic reflecting NASA and NOAA's findings on this: https://www.climatecentral.org/gallery/graphics/the-10-hottest-global-years-on-record.
4. https://www.noaa.gov/news/september-2019-tied-as-hottest-on-record-for-planet.
5. https://www.cbsnews.com/news/earth-just-experienced-its-hottest-october-ever/.
6. https://blogs.scientificamerican.com/eye-of-the-storm/november-2019-earths-second-warmest-november-on-record/.
7. https://www.cbsnews.com/news/australia-man-cooks-roast-pork-in-car-during-heat-wave/.
8. https://climate.copernicus.eu/surface-air-temperature-january-2020.
9. https://www.ncdc.noaa.gov/sotc/global/202002.
10. https://www.nytimes.com/interactive/2018/07/11/climate/summer-nights-warming-faster-than-days-dangerous.html.
11. Ibid.
12. https://www.noaa.gov/new-noaa-study-estimates-future-loss-labor-capacity-climate-warms.
13. Huang, K. Li, X. Liu, X. et al. (2019). Projecting global urban land expansion and heat island intensification through 2050. Environmental Research Letters, 14 (11), p.114037. DOI: 10.1088/1748-9326/ab4b71.
14. See related research from NOAA: https://research.noaa.gov/article/ArtMID/587/ArticleID/2558/Heat-waves-could-increase-substantially-in-size-by-mid-century-says-new-study.
15. NYT had particularly good coverage of these heatwaves: https://www.nytimes.com/2019/07/01/climate/europe-heat-wave.html.
16. Laine, H. Salpakari, J. Looney, E. et al. (2019). Meeting global cooling demand with photovoltaics during the 21st century. Energy & Environmental Science. DOI: 10.1039/C9EE00002J.
17. https://www.rmi.org/wp-content/uploads/2018/11/Global_Cooling_Challenge_Report_2018.pdf.
18. https://www.yaleclimateconnections.org/2019/06/rising-demand-for-air-conditioning-could-make-climate-change-even-worse/.
19. https://www.iea.org/reports/the-future-of-cooling.
20. https://advances.sciencemag.org/content/3/8/e1603322.
21. In his incisive new work *Falter* (2019)
22. https://www.theguardian.com/environment/2019/dec/13/australias-bushfires-have-emitted-250m-tonnes-of-co2-almost-half-of-countrys-annual-emissions.
23. https://www.sciencealert.com/the-amazon-is-burning-at-a-record-rate-and-parts-were-intentionally-set-alight.

24. https://www.nasa.gov/feature/goddard/2019/satellite-data-record-shows-climate-changes-impact-on-fires.

25. https://www.usgs.gov/news/fast-fire-facts-usgs.

26. https://www.nytimes.com/2017/10/10/us/santa-rosa-california-fire.html.

27. Figures courtesy of Kendra Pierre-Louis at NYT: https://www.nytimes.com/2019/08/28/climate/fire-amazon-africa-siberia-worldwide.html.

28. https://twitter.com/SanGasso/status/1154078682509692928.

29. https://www.nasa.gov/feature/goddard/2019/satellite-data-record-shows-climate-changes-impact-on-fires.

30. https://www.reuters.com/article/us-climate-change-permafrost/scientists-amazed-as-canadian-permafrost-thaws-70-years-early-idUSKCN1TJ1XN.

31. https://thehill.com/opinion/energy-environment/456112-greenlands-ice-sheet-wasnt-expected-to-melt-like-this-until-2070.

32. https://www.washingtonpost.com/weather/2019/07/31/greenland-ice-sheet-is-throes-one-its-greatest-melting-events-ever-recorded/.

33. https://www.pnas.org/content/early/2019/07/02/1904822116.

34. Sutherland, D. Jackson, R. Kienholz, J. et al. (2019). Direct observations of submarine melt and subsurface geometry at a tidewater glacier. Science. DOI: 10.1126/science.aax3528.

35. https://advances.sciencemag.org/content/5/6/eaav7266.

36. Read a concise summation of the loss here: https://www.theguardian.com/environment/2019/jun/19/himalayan-glacier-melting-doubled-since-2000-scientists-reveal.

37. https://www.unenvironment.org/resources/publication/global-linkages-graphic-look-changing-arctic.

38. https://www.nature.com/articles/s41467-018-08240-4#ref-CR1.

39. https://www.rochester.edu/newscenter/combat-climate-change-human-activities-more-important-natural-feedbacks-416672/.

40. https://www.telegraph.co.uk/news/2019/10/08/russian-scientists-find-powerful-ever-methane-seep-arctic-ocean/.

41. https://twitter.com/climate_ice/status/494146188514693120?lang=en.

42. https://www.nature.com/articles/s41467-019-12808-z.

43. NYT featured a rundown of all the major cities poised for disappearance: https://www.nytimes.com/interactive/2019/10/29/climate/coastal-cities-underwater.html.

44. Eric Leuliette at the National Oceanic and Atmospheric Administration is cited in the NYT expressing these sentiments here: https://www.nytimes.com/interactive/2016/09/12/science/earth/ocean-warming-climate-change.html.

45. http://dx.doi.org/10.1073/pnas.1901084116.

46. http://science.sciencemag.org/content/363/6430/979.

47. See some good coverage on this in BBC here: https://www.bbc.com/news/science-environment-49255642.

48. Marsooli, R. Lin, N. Emanuel, J. et al. (2019). Climate change exacerbates hurricane flood hazards along US Atlantic and Gulf Coasts in spatially

varying patterns. Nature Communications, 10 (1). DOI: 10.1038/s41467-019-11755-z.

49. https://scienceisessential.org/wp-content/uploads/sites/11/2019/09/Surging_Waters_credits_pages_web.pdf.

50. Zhou, S. Williams, W. Berg, A. et al. (2019). Land-atmosphere feedbacks exacerbate concurrent soil drought and atmospheric aridity. PNAS. DOI: 10.1073/pnas.1904955116.

51. https://www.nytimes.com/2019/03/19/world/europe/uk-water-shortage.html.

52. https://www.nature.com/articles/s41586-019-1822-y.

53. Ibid.

54. https://ocean.si.edu/ocean-life/invertebrates/ocean-acidification.

55. Osborne, E. Thunell, R. Gruber, N. et al. (2019). Decadal variability in twentieth-century ocean acidification in the California Current Ecosystem. Nature Geoscience. DOI: 10.1038/s41561-019-0499-z.

56. https://royalsocietypublishing.org/doi/full/10.1098/rspb.2015.3046.

57. National Geographic carried particularly good coverage of this research here: https://www.nationalgeographic.com/environment/2019/05/ipbes-un-biodiversity-report-warns-one-million-species-at-risk/.

58. https://www.un.org/sustainabledevelopment/blog/2019/05/nature-decline-unprecedented-report/.

59. https://uanews.arizona.edu/story/study-onethird-plant-and-animal-species-could-be-gone-50-years.

60. https://www.sciencedirect.com/science/article/abs/pii/S0006320718313636.

61. https://www.reuters.com/article/us-climatechange-temperatures/evidence-for-man-made-global-warming-hits-gold-standard-scientists-idUSKCN1QE1ZU.

62. An ongoing tracker of environmental roll-backs is available here: https://eelp.law.harvard.edu/regulatory-rollback-tracker/
And here: http://columbiaclimatelaw.com/resources/climate-deregulation-tracker/.

63. According to Coral Davenport, NYT Climate Reporter, via interview with ClimateOne: https://climateone.org/audio/big-climate-stories-2019.

64. https://www.theguardian.com/environment/2019/dec/06/greta-thunberg-says-school-strikes-have-achieved-nothing.

65. https://www.commondreams.org/news/2019/12/06/even-500000-march-madrid-greta-thunberg-warns-climate-movement-has-achieved-nothing.

66. https://time.com/5735388/climate-change-eco-anxiety/.

67. Mach, K. Kraan, C. Adger, W. et al. (2019). Climate as a risk factor for armed conflict. Nature. DOI: 10.1038/s41586-019-1300-6.

68. https://viterbischool.usc.edu/news/2020/01/sea-level-rise-could-reshape-the-united-states-trigger-migration-inland/.

69. https://cires.colorado.edu/news/global-warming-increase-violent-crime-united-states.

70. https://oxfamapps.org/media/press_release/a-person-forced-from-home-every-two-seconds-by-climate-related-disasters-oxfam/.

71. https://www.theatlantic.com/magazine/archive/2018/10/william-vollmann-carbon-ideologies/568309/.

CHAPTER 3: POWERING OUR CIVILIZATIONS
CLEANLY AND SUSTAINABLY

1. See also: Jacobson, M. Delucchi, M. (2011). Providing All Global Energy with Wind, Water, and Solar Power, Part I: Technologies, Energy Resources, Quantities and Areas of Infrastructure, and Materials. Energy Policy, 6 (3). DOI: 10.1016/j.enpol.2010.11.040.
2. https://www.cell.com/one-earth/fulltext/S2590-3322(19)30225-8#%20.
3. See: http://www.epa.gov/ghgemissions/sources-greenhouse-gas-emissions/.
4. From the IEA: http://www.iea.org/statistics/co2emissios/.
5. http://www.witricity.com.
6. Allied Market Research: http://www.alliedmarketresearch.com/wireless-electric-vehicle-charging-market/.
7. www.eroadarlanda.com.
8. For more information, see studies like the below:

   Zaheer, A. Covic, GA. (2016). A comparative study of various magnetic design topologies for a semi-dynamic EV charging application. In: 2016 IEEE 2nd annual southern power electronics conference (SPEC). pp.1–6. http://dx.doi.org/10.1109/SPEC.2016.7846203.

   Stramati, T. Bauer, P. (2013). On-road charging of electric vehicles. In: 2013 IEEE transportation electrification conference and expo (ITEC). pp.1–8. http://dx.doi.org/10.1109/ITEC.2013.6573511.

   Fuller M. (2016). Wireless charging in california: range, recharge, and vehicle electrification. Transport Res Part C: Emerg Technol, 67(Suppl C), pp.343–56. http://dx.doi.org/10.1016/j.trc.2016.02.013 http://www.sciencedirect.com/science/article/pii/S0968090X16000668.

   Chopra, S. Bauer, P. (2013). Driving range extension of EV with on-road contactless power transfer—a case study. IEEE Trans Ind Electron, 60 (1), pp.329–38. http://dx.doi.org/10.1109/TIE.2011.2182015 http://ieeexplore.ieee.org/document/6117077/.

   Covic, G.A. Boys, J.T. (2013). Modern trends in inductive power transfer for transportation applications. IEEE J Emerg Sel Top Power Electron, 1 (1), pp.28–41. http://dx.doi.org/10.1109/JESTPE.2013.2264473 http://ieeexplore.ieee.org/document/6517868/.

9. Based on the Qualcomm HaloTM wireless electric vehicle charging technology (WEVC), Qualcomm Technologies designed and built a wireless DEVC system capable of charging an electric vehicle (EV) dynamically at up to 20 kilowatts at highway speeds.
10. https://www.drawdown.org/solutions/coming-attractions/autonomous-vehicles.
11. https://www.energy.gov/eere/vehicles/downloads/enabling-extreme-fast-charging-technology-gap-assessment.
12. https://www.cell.com/joule/fulltext/S2542-4351(19)30481-7.
13. In October 2019, a Maryland gas station became the first in America to go all-electric for including only EV charging stations.

https://www.cnbc.com/2019/09/26/first-gas-station-to-ditch-oil-for-electric-vehicle-charging-now-open.html.

14. https://www.forbes.com/sites/mikescott/2019/06/10/electric-models-to-dominate-car-sales-by-2040-wiping-out-13m-barrels-a-day-of-oil-demand/.

15. https://www.theguardian.com/world/2019/may/03/amsterdam-ban-petrol-diesel-cars-bikes-2030.

16. https://www.reuters.com/article/us-norway-autos/norways-electric-cars-zip-to-new-record-almost-a-third-of-all-sales-idUSKCN1OW0YP.

17. https://www.bbc.com/news/world-europe-48668791.

18. https://www.ucsusa.org/about/news/heavy-duty-truck-electrification-critical-and-viable-climate-solution-california-says.

19. https://www.fortune.com/2019/12/23/rivian-fund-raising-amazon/.

20. Good coverage and commentary here: https://nationalpost.com/news/world/will-new-electric-ferries-lead-the-way-in-cutting-maritime-pollution/.

21. https://www.forbes.com/sites/jamesellsmoor/2019/08/18/the-worlds-largest-electric-ferry-has-completed-its-maiden-voyage/.

22. For a discussion of some case studies, despite the higher up front cost, see: https://www.insideclimatenews.org/news/14112019/electric-bus-cost-savings-health-fuel-charging/.

23. https://freewiretech.com/.

24. https://www.volkswagenag.com/en/news/stories/2019/12/volkswagen-lets-its-charging-robots-loose.html.

25. http://www.exeter.ac.uk/news/research/title_783082_en.html.

26. Zonghai Chen, J. Ma, Z. Pan, F. et al. (2016). The role of nanotechnology in the development of battery materials for electric vehicles. Nature Nanotechnology, 11, pp.1031-1038.

27. https://www.bloomberg.com/opinion/articles/2019-04-12/electric-vehicle-battery-shrinks-and-so-does-the-total-cost.

28. https://www.sciencedirect.com/science/article/pii/S0360128516300442.

29. See the post in their News Room section: Engine design in the automotive industry: optimized wear due to modern surface coating process. http://appliednanosurfaces.com/newsroom/.

30. https://www.woodmac.com/reports/gas-markets-the-rise-of-the-electric-car-how-will-it-impact-oil-power-and-metals-10987/ https://www.woodmac.com/reports/gas-markets-the-rise-of-the-electric-car-how-will-it-impact-oil-power-and-metals-10987/.

31. https://www.nytimes.com/2019/06/13/opinion/letters/travel-climate-change.html.

32. https://www.neste.com/companies/products/renewable-fuels/neste-my-renewable-jet-fuel/.

33. https://www.marketwatch.com/story/this-magic-word-mayh-convince-airline-passengers-to-spend-more-money-to-protect-the-environment-2019-10-07/.

34. https://www.eviation.co/alice/.

35. https://www.airbus.com/innovation/future-technology/electric-flight.html.

36. http://www.solarroadways.com/.

37. http://www.solarroadways.com/Blog/Show?b=1.
38. You'll need to translate for the details: https://www.pv-magazine. de/2019/11/19/warum-deutschlands-erster-solarradweg-vornehmlich-nur- noch-anwaelte-bewegt/.
39. Venugopal, P. Shekhar, A. Visser, E. Scheele, N. Chandra, G. (2018). Roadway to self-healing highways with integrated wireless electric vehicle charging and sustainable energy harvesting technologies. Applied Energy, 212, pp.1226-1239. https://doi.org/10.1016/j.apenergy.2017.12.108.
40. http://www.xinhuanet.com/english/2019-08/16/c_138314484.htm.
41. https://www.bloomberg.com/news/features/2018-04-11/the-solar-highway- that-can-recharge-electric-cars-on-the-move.
42. Hashem, A. Leanna, S.R. (2001). Characterizing the fabric of the urban environment: a case study of metropolitan Chicago, Illinois.
43. Cited within a great article on WEF, even if the company itself remains elusive: (2018). Tokyo Plans to install solar roads ahead of the 2020 Olympics. World Economic Forum. https://www.weforum.org/ agenda/2018/06/tokyo-announces-plan-to-install-solar-roads-in-time-for- 2020-olympics/.
44. www.pavegen.com.
45. http://www.platio.cc/en/home/en/products_/.
46. Efthymiou, C. Santamouris, M. Kolokotsa. D. et al. (2016). Development and testing of photovoltaic pavement for heat island mitigation. Sol Energy, 130, pp.148–60.
47. Ma, T. Yang, H. Gua, W. Lia, Z. Yana, S. (2019). Development of walkable photovoltaic floor tiles used for pavement. https://doi.org/10.1016/j. enconman.2019.01.035.
48. Cui, Y. Wang, Y. Bergqvist, J. et al. (2019). Wide-gap non-fullerene acceptor enabling high-performance organic photovoltaic cells for indoor applications. Nature Energy, 4 (9), p.768 DOI: 10.1038/s41560-019-0448-5.
49. Michaels, H. Rinderle, M. Freitag, R. et al. (2020). Dye-sensitized solar cells under ambient light powering machine learning: towards autonomous smart sensors for the internet of things. Chemical Science. DOI: 10.1039/ C9SC06145B.
50. Röhr, J. Lipton, J. Kong, J. et al. (2020). Efficiency Limits of Underwater Solar Cells. Joule. DOI: 10.1016/j.joule.2020.02.005.
51. https://www.weforum.org/agenda/2019/03/japan-is-the-world-leader-in- floating-solar-power/.
52. https://news.trust.org/item/20190208102615-bye4f/.
53. (2018). Mega floating solar panel system to be ready in 2019. The Straits Times. https://www.straitstimes.com/business/mega-floating-solar-panel- system-to-be-ready-in-2019.
54. https://news.trust.org/item/20190208102615-bye4f/.
55. Nandjou, T. (2019). A thermally synergistic photo-electrochemical hydrogen generator operating under concentrated solar irradiation. Nature Energy. DOI: 10.1038/s41560-019-0373-7.

56. http://www.businesskorea.co.kr/news/articleView.html?idxno=34083.
57. https://www.woodmac.com/our-expertise/focus/Power--Renewables/floating-solar-2019/.
58. https://www.nrel.gov/news/press/2018/nrel-details-great-potential-for-floating-pv-systems.html.
59. Liu, Z. Sofia, S. Laine, H. et al. (2020). Revisiting thin silicon for photovoltaics: a technoeconomic perspective. Energy & Environmental Science, 13 (1), p.12 DOI: 10.1039/C9EE02452B.
60. A good writeup on perovskite's potential as a solar energy material appeared on WEF near the end of 2019. I recommend it: https://www.weforum.org/agenda/2019/08/this-new-solar-tech-can-be-printed-on-paper-or-woven-into-fabric/.
61. https://www.oxfordpv.com/news/oxford-pv-perovskite-solar-cell-achieves-28-efficiency.
62. See commentary in *Nature* for elaboration: https://www.nature.com/articles/d41586-019-01985-y.
63. (2019). Broadband antireflective and superhydrophobic coatings for solar cells. Materials Today Energy, 12, pp. 348-355. https://doi.org/10.1016/j.mtener.2019.03.006.
64. https://news.rice.edu/2019/07/12/rice-device-channels-heat-into-light/.
65. https://www.uq.edu.au/news/article/2020/02/solar-technology-breakthrough-uq.
66. Alam, M. Khan, M. (2019). Shockley–Queisser triangle predicts the thermodynamic efficiency limits of arbitrarily complex multijunction bifacial solar cells. Proceedings of the National Academy of Sciences, 116 (48), p.23966 DOI: 10.1073/pnas.1910745116.
67. See: https://spectrum.ieee.org/energy/renewables/the-dawn-of-solar-windows.
68. Kang, S. Kim, J. Jeong, M et al. (2019). Stretchable and colorless freestanding microwire arrays for transparent solar cells with flexibility. Light: Science & Applications. DOI: 10.1038/s41377-019-0234-y.
69. http://www.soliculture.com/lumo-technology/.
70. Liu, Y. et al. (2019). Unraveling sunlight by transparent organic semiconductors toward photovoltaic and photosynthesis. ACS Nano, DOI: 10.1021/acsnano.8b08577.
71. Kashyap, V. Sakunkaewkasem, S. Jafari, P. et al. (2019). Full Spectrum Solar Thermal Energy Harvesting and Storage by a Molecular and Phase-Change Hybrid Material. Joule. DOI:https://doi.org/10.1016/j.joule.2019.11.001.
72. http://news.mit.edu/2019/increase-solar-cell-output-photon-2-electron-0703.
73. (2017). China's world-beating solar farm is almost as big as Macau, Nasa satellite images reveal. *South China Morning Post*. https://www.scmp.com/news/china/society/article/2073747/powerful-images-worlds-largest-solar-energy-farms-are-china.
74. https://www.iea.org/wei2019/.
75. Nagendran, S. (2017). 4 Charts Explaining Latin America's Impending Solar Boom. Greentech Media, https://www.greentechmedia.com/articles/read/explaining-latin-americas-impending-solar-boom1.

76. As reported in the *LA Times*: https://www.latimes.com/business/la-fi-solar-power-new-homes-20181206-story.html.

77. See: https://environmentamerica.org/sites/environment/files/reports/WEB_AME_Shining-Cities_2019_040919-v1.pdf.

78. Hernandez, R. Armstrong, A. Burney, J. (2019). Techno–ecological synergies of solar energy for global sustainability. Nature Sustainability, 2 (7), p.560 DOI: 10.1038/s41893-019-0309-z.

79. Rohrig, K. Berkhout, V. Callies, D. et al. (2019). Powering the 21st century by wind energy—Options, facts, figures. Applied Physics Reviews, 6 (3). DOI: 10.1063/1.5089877.

80. https://eerscmap.usgs.gov/uswtdb/.

81. https://www.bnnbloomberg.ca/first-big-u-s-offshore-wind-farm-offers-1-4-billion-to-customers-1.1117584.

82. https://www.cnbc.com/2019/03/07/uk-says-offshore-wind-will-provide-one-third-of-electricity-by-2030.html.

83. https://gwec.net/wp-content/uploads/2019/04/GWEC-Global-Wind-Report-2018.pdf.

84. https://denmark.dk/innovation-and-design/clean-energy.

85. https://www.iea.org/geco/.

86. https://gwec.net/wp-content/uploads/2019/04/GWEC-Global-Wind-Report-2018.pdf.

87. Enevoldsen, P. Permien, F. Bakhtaoui, I. et al. (2019). How much wind power potential does Europe have? Examining European wind power potential with an enhanced socio-technical atlas. Energy Policy, 132, p.1092 DOI: 10.1016/j.enpol.2019.06.064.

88. See the university's PR post here: http://news.cornell.edu/stories/2020/02/quadrupling-turbines-us-can-meet-2030-wind-energy-goals.

89. Ibid.

90. https://www.genewsroom.com/press-releases/power-generated-ge-haliade-x-12-mw-prototype-rotterdam-be-bought-eneco-utility.

91. Lee, J. Kessler, S. Wardle, B. (2020). Void-Free Layered Polymeric Architectures via Capillary-Action of Nanoporous Films. Advanced Materials Interfaces. DOI: 10.1002/admi.201901427.

92. https://x.company/projects/makani/.

93. Silva-Leon, J. Cioncolini, A. Nabawy, M. et al. (2019). Simultaneous wind and solar energy harvesting with inverted flags. Applied Energy, 239. DOI: 10.1016/j.apenergy.2019.01.246.

94. https://skysails-power.com/advantages.html.

95. https://www.irena.org/publications/2019/Oct/Future-of-wind.

96. https://sciencebasedmedicine.org/health-effect-of-wind-turbines/.

97. A very insightful Scientific American piece describes the study and the phenomenon here: https://www.scientificamerican.com/article/the-worlds-winds-are-speeding-up/. Original study described by the university's PR post: https://engineering.princeton.edu/news/2019/11/18/boosting-wind-farmers-global-winds-reverse-decades-slowing-

CHAPTER 4: BROADENING OUR ENERGY PORTFOLIO

1. Via interview with *Forbes*: https://www.forbes.com/sites/cognitiveworld/2019/09/23/un-climate-summit-getting-to-drawdown-with-microsoft-ai-interview/.

2. Raman, A. Shanhui Fan, W. (2019). Generating Light from Darkness. Joule, DOI: 10.1016/j.joule.2019.08.009.

3. https://pubs.acs.org/doi/10.1021/acsphotonics.9b00679.

4. One of the many new studies getting bioprospectors excited: Banerjee, A. Banerjee, C. Negi, Chang, JS. Shukla, P. (2018). Improvements in algal lipid production: a systems biology and gene editing approach. Crit Rev Biotechnol. 38(3), pp.369-385. DOI: 10.1080/07388551.2017.1356803.

5. For example, see: Verruto, A. Aqui, M. Soriaga, LB. et al. (2017). Lipid production in Nannochloropsis gaditana is doubled by decreasing expression of a single transcriptional regulator. Nat Biotechnol, 35(7), pp.647-652. DOI: 10.1038/nbt.3865.

6. http://www.photofuel.eu/home.php.

7. For a good rundown of the research, see: https://sciencetrends.com/microalgae-the-green-mines-for-fueling-the-future-and-mitigating-heavy-metals/.

8. Gunther, G. Reichelstein, S. (2019). Economics of converting renewable power to hydrogen. Nature Energy. DOI: 10.1038/s41560-019-0326-1.

9. https://www.cnbc.com/2019/02/21/musk-calls-hydrogen-fuel-cells-stupid-but-tech-may-threaten-tesla.html.

10. https://www.northerngasnetworks.co.uk/event/H21-launches-national/.

11. https://www.reuters.com/article/us-france-bicycles-hydrogen/french-startup-launches-hydrogen-powered-bicycles-idUSKBN1F52AP.

12. https://www.youtube.com/watch?v=_U_dHZZ3vxQ.

13. See Engelbrecht and Happe, in Methods in Enzymology (2018). https://www.sciencedirect.com/science/article/pii/S007668791830418X.

14. https://www.pnas.org/content/116/25/12212.

15. According to researchers via interview with NBC, that is: https://www.nbcnews.com/mach/science/how-floating-solar-farms-could-make-fuel-help-solve-climate-ncna1020336.

16. Kuang, Y. Kenney, M. Meng, Y. et al. (2019). Solar-driven, highly sustained splitting of seawater into hydrogen and oxygen fuels. Proceedings of the National Academy of Sciences. DOI: 10.1073/pnas.1900556116.

17. https://acee.princeton.edu/acee-news/researchers-use-sunlight-to-pull-hydrogen-from-wastewater/. Lu, L. Vakki, W. Anguiar, A. et al. (2019). Unbiased solar H2 production with current density up to 23 mA cm-2 by Swiss-cheese black Si coupled with wastewater bioanode. Energy & Environmental Science, 12 (3), p.1088. DOI: 10.1039/c8ee03673j.

18. https://www.ethz.ch/en/news-and-events/eth-news/news/2019/06/pr-solar-mini-refinery.html.

19. It is chemically slightly different from the kerosene we're used to. There are a few different ways to "green" kerosene for aviation. See a useful discussion

of this here: https://www.chemeurope.com/en/news/1160637/industrial-consortium-on-the-way-to-green-kerosene.html.

20. https://ethz.ch/en/news-and-events/eth-news/news/2019/06/pr-solar-mini-refinery.html.

21. Hannon, J. Lynd, L. Andrade, O. et al. (2019). Technoeconomic and life-cycle analysis of single-step catalytic conversion of wet ethanol into fungible fuel blendstocks. Proceedings of the National Academy of Sciences. DOI: 10.1073/pnas.1821684116.

22. PR post from the University of Utah is where I most recently noticed the stat: https://unews.utah.edu/beat-the-heat.

23. https://www.hku.hk/press/news_detail_20140.html.

24. Michael G. Stanford, John T. Li, Yieu Chyan, Zhe Wang, Winston Wang, James M. Tour. Laser-Induced Graphene Triboelectric Nanogenerators. ACS Nano, 2019; DOI: 10.1021/acsnano.9b02596.

25. Sargolzaeiaval, Y. Ramesh, V. Neumann, T. et al. (2020). Flexible thermoelectric generators for body heat harvesting – Enhanced device performance using high thermal conductivity elastomer encapsulation on liquid metal interconnects. Applied Energy, 262. DOI: 10.1016/j.apenergy.2019.114370. See also another design from Japanese scientists here: Shibata, T. Iwaizumi, H. Fukuzumi, Y. et al. (2020). Energy harvesting thermocell with use of phase transition. Scientific Reports, 10 (1). DOI: 10.1038/s41598-020-58695-z.

26. https://www.cnn.com/2019/02/06/motorsport/morocco-solar-farm-formula-e-spt-intl/index.html.

27. Tembhurne, S. Nandjou, F. Haussener, S. (2019). A thermally synergistic photo-electrochemical hydrogen generator operating under concentrated solar irradiation. Nature Energy, 4, pp.399–407. DOI: 10.1038/s41560-019-0373-7.

28. Meder, F. Must, I. Sadeghi, A. et al. (2018). Energy Conversion at the Cuticle of Living Plants. Advanced Functional Materials, DOI: 10.1002/adfm.201806689.

29. http://onlinelibrary.wiley.com/wol1/doi/10.1002/anie.201602114/abstract.

30. https://www.cityu.edu.hk/media/news/2020/02/06/cityu-new-droplet-based-electricity-generator-drop-water-lights-100-small-led-bulbs.

31. https://www.nature.com/articles/s41586-020-2010-9.

CHAPTER 5: KEEPING THE LIGHTS ON

1. https://ourworldindata.org/grapher/renewables-share-electricity-production/.

2. https://www.weforum.org/agenda/2019/10/australia-renewable-energy-hydropower-national-park/.

3. https://www.nbcnews.com/mach/science/tidal-energy-pioneers-see-vast-potential-ocean-currents-ebb-flow-ncna981341.

4. https://news.berkeley.edu/2014/01/28/seafloor-carpet-catches-waves-to-harness-energy/.

5. Reported in Reuters: https://mena.projects.thomsonreuters.com/newsDetails.html?newsId=ZAWYA20190529080354#/.
6. https://www.turbulent.be/.
7. https://www.businessinsider.com/whirlpool-turbine-water-energy-turbulent-belgian-green-energy-2018-2.
8. https://www.oist.jp/news-center/news/2017/9/20/sustainable-future-powered-sea.
9. https://www.drawdown.org/solutions/coming-attractions/solid-state-wave-energy.
10. Ibid.
11. https://blueeconomycrc.com.au/about/.
12. Calculated with the help of the DoE: http://www.energy.gov/eere/articles/how-much-power-1-gigawatt.
13. https://www.energy.gov/eere/articles/5-things-know-about-geothermal-power.
14. Ibid.
15. https://www.energy.gov/eere/articles/how-much-power-1-gigawatt.
16. https://e360.yale.edu/features/aiming-for-climate-targets-germany-taps-its-geothermal-potential.
17. https://e360.yale.edu/features/aiming-for-climate-targets-germany-taps-its-geothermal-potential.
18. https://e360.yale.edu/features/aiming-for-climate-targets-germany-taps-its-geothermal-potential.
19. (2019). South Korea accepts geothermal plant probably caused destructive quake. *Nature.* https://www.nature.com/articles/d41586-019-00959-4.
20. According to Thomas Öström, Climeon's CEO, via interview: https://www.marinelog.com/news/viking-line-says-climeon-system-will-bring-major-co2-emissions-cuts/.
21. Ibid.
22. According to Dandelion's own recount: https://dandelionenergy.com/blog/introducing-dandelion
23. https://enerdrape.com/.
24. https://spectrum.ieee.org/energy/renewables/altarock-energy-melts-rock-with-millimeter-waves-for-geothermal-wells.
25. https://spectrum.ieee.org/energywise/energy/renewables/is-google-going-underground-with-hypersonic-tech.
26. http://news.mit.edu/2017/3-d-graphene-strongest-lightest-materials-0106.
27. https://www.youtube.com/watch?v=YY7f1t9y9a0.
28. https://www.bnnbloomberg.ca/elon-musk-s-outback-battery-sparks-new-projects-after-promising-run-1.1177863.
29. https://ambri.com/technology/.
30. https://www.cnbc.com/2019/03/15/battery-start-ups-are-raising-millions-in-the-battle-to-crush-tesla.html.
31. https://www.woodmac.com/news/editorial/the-future-for-green-hydrogen/.
32. https://energyvault.ch/news/.
33. https://iiasa.ac.at/web/home/about/news/200219-seasonal-pumped-storage.html.

34. https://www.wsset.org/docs/articles/deep-storage.pdf.
35. See: https://www.ncbi.nlm.nih.gov/pubmed/21413685
    And: http://techfinder.stanford.edu/technologies/S10-123_mixing-entropy-battery.
36. https://greensync.com/.
37. For that stat and more reliable exposition on the positives of nuclear over solar and wind, see Michael Shellenberger's piece in *Forbes* here and follow through to his website: https://www.forbes.com/sites/michaelshellenberger/2018/05/08/we-dont-need-solar-and-wind-to-save-the-climate-and-its-a-good-thing-too/.
38. See the team's interview in *Forbes*: https://www.forbes.com/sites/christopherhelman/2019/05/21/the-new-nuclear-how-a-600-million-fusion-energy-unicorn-plans-to-beat-solar/#39c03aa3629e.
39. https://www.moltexenergy.com/ourbreakthrough/.
40. https://www.iea.org/reports/tracking-power-2019/nuclear-power.
41. https://www.drawdown.org/drawdown-framework/drawdown-review-2020.
42. I recommend a write-up here discussing the Allam Cycle for additional clarity: https://www.powermag.com/inside-net-power-gas-power-goes-supercritical/.
43. Another good write-up on Allam Cycle plants and efficiency ratios here: https://www.vox.com/energy-and-environment/2018/6/1/17416444/net-power-natural-gas-carbon-air-pollution-allam-cycle.

CHAPTER 6: INDUSTRY WITHOUT EMISSIONS

1. https://www.eia.gov/tools/faqs/faq.php?id=99&t=3.
2. Plants draw in water through their roots and fix chemicals from the air through leaf components. As such, they will draw in a number of chemicals unintended for use. If one of those chemicals—say, a compound like nitroaromatics, used in landmines and other explosives—runs into other structures placed into plants, researchers can elicit specific reactions between the two; reactions which then trigger other events. In this case, carbon nanotubes embedded within plant leaves can emit a fluorescent signal which is then read with an infrared camera, which then pings a user. http://www.news.mit.edu/2016/nanobionic-spinach-plants-detect-explosives-1031/.
3. https://www.c2es.org/content/u-s-emissions/.
4. You'll find this question in her TEDxTalk video and via interview in Susan Hockfield's latest work The Age of Living Machines.
5. Via Kelly's interview with TechCrunch: https://techcrunch.com/2019/19/ginkgo-bioworks-dev-shop-for-genetic-programming-is-now-worth-4-billion/.
6. Ibid.
7. As per their site: https://cariboubio.com/application-areas/industrial-biotech.
8. For example, see: https://www.cell.com/molecular-plant/fulltext/S1674-2052(18)30374-5.
9. https://www.uni-bonn.de/news/055-2020.

10. https://www.ted.com/talks/angela_belcher_using_nature_to_grow_batteries/.
11. http://advances.sciencemag.org/content/5/1/eaat5664.
    Additional exposition: https://www.forbes.com/sites/jeffkart/2019/01/25/new-technique-could-put-electricity-producing-bacteria-to-work/#3b1926203295.
12. Ngo-Duc, T. Plank, J. Chen, G. et al. (2018). M13 bacteriophage spheroids as scaffolds for directed synthesis of spiky gold nanostructures. Nanoscale, 10 (27), p. 13055-13063. DOI: 10.1039/C8NR03229G.
13. https://ucrtoday.ucr.edu/55006.
14. Ding, Y. Bertram, J. Eckert, C. et al. (2019). Light-driven renewable biochemical synthesis using quantum dot-bacteria nano-biohybrids. Journal of the American Chemical Society. DOI: 10.1021/jacs.9b02549.
15. For a rundown of our progress, I recommend a secondary source: https://www.greenbiz.com/article/aluminum-can-americas-most-successful-recycling-story-youve-never-heard.
16. https://michelinmedia.com/michelin-uptis/.
17. Gebler, M. Anton, J. Uiterkamp, et al. (2014). A global sustainability perspective on 3D printing technologies. Energy Policy, 74, pp.158-167. DOI: 10.1016/j.enpol.2014.08.033.
18. https://www.drawdown.org/solutions/materials/alternative-cement.
19. Drawdown lists 40 percent of cement's emissions in the energy-side of refinement, aside from the 60 percent which results from decarbonization.
20. https://mushroompackaging.com/.

CHAPTER 7: FEEDING THE POST-CARBON WORLD

1. See: https://data.worldbank.org/indicator/EN.ATM.CO2E.KT.
2. http://www.fao.org/news/story/en/item/197623/icode/.
3. Xu, R. Zhao, Q. Coelho, M. et al. (2019). The association between heat exposure and hospitalization for undernutrition in Brazil during 2000–2015: A nationwide case-crossover study. PLOS Medicine, 16 (10). DOI: 10.1371/journal.pmed.1002950.
4. http://ipcc.ch/2019/08/08/land-is-a-critical-resource_srccl/.
5. http://smartmachines.bluerivertechnology.com/.
6. https://www.naio-technologies.com/experts-robotique-agricole/.
7. From researchers via interview: https://www.telegraph.co.uk/global-health/climate-and-people/scientists-develop-super-spud-bid-prevent-stunting/.
8. Via interview with WEF: https://www.weforum.org/agenda/2019/07/these-tech-start-ups-are-changing-what-it-means-to-farm/.
9. https://science.sciencemag.org/content/365/6452/422?_ga=2.220387422.739930905.1565038064-1966656220.1563490071.
10. South, P. et al. (2019). Synthetic glycolate metabolism pathways stimulate crop growth and productivity in the field. Science. DOI: 10.1126/science.aat9077.
11. https://emails.illinois.edu/newsletter/202480.html.

12. Shen, B. Wang, L. Lin, X. et al. (2019). Engineering a New Chloroplastic Photorespiratory Bypass to Increase Photosynthetic Efficiency and Productivity in Rice. Molecular Plant. DOI: 10.1016/j.molp.2018.11.013.

13. Schuler, M. Mantegazza, O. Weber, A. (2016). Engineering C4 photosynthesis into C3 chassis in the synthetic biology age. Special Issue: Synthetic Biology for Basic and Applied Plant Research, 87(1). pp. 51-65. DOI: https://doi.org/10.1111/tpj.13155.

14. Malone, L. Qian, P. Mayneord, G. et al. (2019). Cryo-EM structure of the spinach cytochrome b6f complex at 3.6 Å resolution. *Nature*. DOI: 10.1038/s41586-019-1746-6.

15. Baulcombe, D. Dunwell, J. Jones, J. et al. (2014). GM Science Update. A report for the Council for Science and Technology, March. https://www.gov.uk/government/publications/genetic-modification-gm-technologies.

16. Tripathi, L. Atkinson, H. Roderick, H. et al. (2017). Genetically engineered bananas resistant to Xanthomonas wilt disease and nematodes. Food and Energy Security, 6 (2), pp.37–47. DOI: 10.1002/fes3.101. See also newer research: https://www.nature.com/articles/s42003-019-0288-7.

17. Sovk, S. Lemmon, Z. Oved, M. et al. (2017). Bypassing Negative Epistasis on Yield in Tomato Imposed by a Domestication Gene. Cell, 169 (6), pp.1142–1155. DOI:https://doi.org/10.1016/j.cell.2017.04.032.

18. https://geneticliteracyproject.org/2017/10/11/green-technology-disease-resistant-gmo-tomato-eliminate-need-copper-pesticides-double-yields-blocked-public-fears/.

19. Zsögön, A. Cermák, T. Rezende, E. et al. (2018). De novo domestication of wild tomato using genome editing. Nature Biotechnology, 36, pp.1211–1216. DOI: 10.1038/nbt.4272.

20. For a discussion of the research from an editorial standpoint from *Nature*: https://www.nature.com/articles/nature.2016.19754.

21. https://massivesci.com/articles/cassava-crispr/.

22. Kost, T. Gessler, C. Jansch, M. et al. (2015). Development of the First Cisgenic Apple with Increased Resistance to Fire Blight. PLOS ONE. https://doi.org/10.1371/journal.pone.0143980.

23. https://www.foodbusinessnews.net/articles/10892-bioengineering-of-wheat-still-faces-significant-challenges.
    And: Sanchez-Leon, S. Gil-Humanes, J. Ozuna, C. et al. (2018). Low-gluten, nontransgenic wheat engineered with CRISPR/Cas9. *Plant Biotechnology Journal*, 16 (4), pp.902-910. DOI: https://doi.org/10.1111/pbi.12837.

24. Caine, R. Yin, X. Sloan, J. et al. (2019). Rice with reduced stomatal density conserves water and has improved drought tolerance under future climate conditions. New Phytologist, 221 (1), pp.371-384. DOI: 10.1111/nph.15344.

25. Azad, M. Amin, L. Marzuki, N. (2014). Gene Technology for Papaya Ringspot Virus Disease Management. Scientific World Journal. DOI: 10.1155/2014/768038.

26. Zhang, R. Liu, J. Chai, Z. et al. (2019). Generation of herbicide tolerance

traits and a new selectable marker in wheat using base editing. Nature Plants, 5, pp.480–485. DOI: 10.1038/s41477-019-0405-0.

27. Morris, J. Hawthorne, K. Hotze, T. et al. (2008). Nutritional impact of elevated calcium transport activity in carrots. PNAS, 105 (5), pp.1431-1435. DOI: https://doi.org/10.1073/pnas.0709005105.

28. Nunes, A. Kalkmann, D. Aragao, F. (2009). Folate biofortification of lettuce by expression of a codon optimized chicken GTP cyclohydrolase I gene. https://doi.org/10.1007/s11248-009-9256-1.

29. Good write-up on ETH Zurich's website here: https://ethz.ch/en/news-and-events/eth-news/news/2017/08/multi-nutrient-rice-against-malnutrition.html.

30. Foo, M. Gherman, I. Zhang, P. (2018). A Framework for Engineering Stress Resilient Plants Using Genetic Feedback Control and Regulatory Network Rewiring. ACS Synthetic Biology, 7 (6), p.1553 DOI: 10.1021/acssynbio.8b00037.

31. Klümper, W., Qaim, M. (2014). A meta-analysis of the impacts of genetically modified crops. PLoS One 9 (11). https://doi.org/10.1371/journal.pone.0111629. e111629.

32. For a great write-up on Bt eggplant, rather than just the academic article, you can check out this editorial on Cornell's site: https://news.cornell.edu/stories/2018/07/bt-eggplant-improving-lives-bangladesh.

33. https://fao.org/save-food/resources/keyfindings/en/.

34. https://cariboubio.com/application-areas/agricultural-biotech.

35. From Michigan State University's assessment: https://canr.msu.edu/resources/management_of_nitrogen_fertilizer_to_reduce_nitrous_oxide_emissions_from_fi.

36. https://insideclimatenews.org/news/11092019/nitrous-oxide-climate-pollutant-explainer-greenhouse-gas-agriculture-livestock.

37. Thompson, R. Lassaletta, L. Patra, P. et al. (2019). Acceleration of global N2O emissions seen from two decades of atmospheric inversion. Nature Climate Change. DOI: 10.1038/s41558-019-0613-7.

38. Zhou, X. Passow, F. Rudek, J. et al. (2019). Estimation of methane emissions from the U.S. ammonia fertilizer industry using a mobile sensing approach. Elem Sci Anth, 7 (1), p.19 DOI: 10.1525/elementa.358.

39. It's worth checking out their website if this sounds complicated (and I realize that it does): https://www.kulabio.com/.

40. https://news.harvard.edu/gazette/story/2018/01/harvards-bionic-leaf-could-help-feed-the-world/.

41. https://www.pivotbio.com/our-science.

42. https://www.indigoag.com/for-consumers.

43. See the PR post at MIT here: http://news.mit.edu/2019/coated-seeds-agriculture-marginal-lands-1125.

44. https://journals.plos.org/plosone/article?id=10.1371/journal.pone.0228305.

45. https://www.wired.com/story/gene-editing-food-climate-change/.

46. https://aerofarms.com/technology/.

47. https://aerofarms.com/environmental-impact/.

48. https://www.smithsonianmag.com/innovation/dubai-will-be-home-to-worlds-biggest-vertical-farm-180969655/.
49. https://www.agricool.co/fr/faq.
50. https://press.nature.com/rapid-customization-of-solanaceae-fruit-crops-for-urban-agricult/17506924.
51. https://www.nature.com/articles/s41596-018-0072-z.
    See also: https://www.nature.com/articles/s41587-019-0152-9.
52. https://www.gatesnotes.com/Energy/My-plan-for-fighting-climate-change.
53. https://www.technologyreview.com/lists/technologies/2019/.
54. On Jon's blog: https://globalecoguy.org/farming-our-way-out-of-the-climate-crisis-c235e1aaff8d.
55. http://science.sciencemag.org/content/360/6392/987.
56. https://www.ncbi.nlm.nih.gov/pmc/articles/PMC3382952/.
57. https://www.thelancet.com/journals/lanplh/article/PIIS2542-5196(19)30245-1/fulltext.
58. https://www.ipcc.ch/report/srccl/.
59. https://www.drawdown.org/solutions/coming-attractions/cowwalks-beach.
60. https://www.theguardian.com/environment/2019/jun/12/most-meat-in-2040-will-not-come-from-slaughtered-animals-report.
61. Kinley, R. Nys, R. Vucko, M. et al. (2016). The red macroalgae Asparagopsis taxiformis is a potent natural antimethanogenic that reduces methane production during in vitro fermentation with rumen fluid. Animal Production Science, 56 (3), p.282. DOI: 10.1071/AN15576.
62. https://advances.sciencemag.org/content/5/7/eaav8391.
63. https://impossiblefoods.com/mission/2019impact/.
64. https//beyondmeat.com/about/.
65. https://www.mosameat.com/faq.
66. https://www.theguardian.com/environment/2019/jun/12/most-meat-in-2040-will-not-come-from-slaughtered-animals-report.
67. https://www.theguardian.com/environment/2019/oct/07/wheres-the-beef-248-miles-up-as-first-meat-is-grown-in-a-space-lab/.
68. https://www.nytimes.com/2019/08/02/science/lab-grown-milk.html.
69. Ibid.
70. https://www.youtube.com/watch?v=BQHFcTdDlRQ.
71. https://www.frontiersin.org/articles/10.3389/fsufs.2019.00005/full.
72. See their original write-up on CNN: https://www.cnn.com/2019/07/16/us/solar-foods-protein-from-air-trnd/index.html.
73. According to a Solar Foods analysis: https://solarfoods.fi/.
74. Cited via interview: https://www.fastcompany.com/90428522/the-newest-fake-meat-is-made-from-thin-air.
75. https://www.ipcc.ch/report/srccl/.

CHAPTER 8: SMART HOMES, SMARTER CITIES

1. https://oceanix.org/net-zery-energy/.
2. https://www.epa.gov/ghgemissions/sources-greenhouse-gas-emissions.
3. https://www.c2es.org/content/home-energy-use/.
4. https://www.news.mit.edu/2019/aerogel-passive-heat-sunlight-0702.
5. https://www.drawdown.org/solutions/buildings-and-cities/smart-glass.
6. For example: Gayathri, P. Shaiju, S. Remya, R. Biswapriya, D. (2018). Hydrated tungsten oxide nanosheet electrodes for broadband electrochromism and energy storage. Materials Today Energy, 10, pp. 380–387. https://doi.org/10.1016/j.mtener.2018.09.006.
7. See: Hadeebllah, O. Garlisia, C. Al-Alia, K. Azarb, E. Palmisanoa, G. (2019). Combined photocatalytic properties and energy efficiency via multifunctional glass. Journal of Environmental Chemical Engineering 7. https://doi.org/10.1016/j.jece.2019.102980.
8. Xin-Hao Li et al. Broadband Light Management with Thermochromic Hydrogel Microparticles for Smart Windows. Joule, 2018.
9. See the coverage of Berkeley labs' work here: https://newscenter.lbl.gov/2016/02/25/berkeley-lab-scientists-developing-paint-on-coating-for-energy-efficient-windows/.
10. Petersen, A. Hofmann, A. Fillols, M. et al. (2019). Solar Energy Storage by Molecular Norbornadiene–Quadricyclane Photoswitches: Polymer Film Devices. Advanced Science. DOI: 10.1002/advs.201900367.
11. https://www.chalmers.se/en/departments/chem/news/Pages/Window-film-could-even-out-the-indoor-temperature-using-solar-energy.aspx.
12. https://www.nature.com/articles/nenergy2017143.
13. https://www.technologyreview.com/s/608840/a-material-that-throws-heat-into-space-could-soon-reinvent-air-conditioning/.
14. Zhou, L. Song, H. Liang, J. et al. (2019). A polydimethylsiloxane-coated metal structure for all-day radiative cooling. Nature Sustainability, 2019 DOI: 10.1038/s41893-019-0348-5.
15. https://www.cnbc.com/2019/05/15/phononic-has-a-new-take-on-cooling-and-a-new-partnership-with-unilever.html.
16. https://iopscience.iop.org/article/10.1088/1748-9326/ab28b0.
17. https://www.cyclotronroad.org/cypris-materials.
18. https://www.weforum.org/agenda/2019/05/the-copenhagen-effect-how-europe-can-become-heat-efficient/.
19. https://www.popularmechanics.com/science/green-tech/a28848192/water-harvester-turns-air-into-water/.
20. https://www.xprize.org/articles/waxp-grand-prize-winner.
    Or see a video of the team's device here: https://www.youtube.com/watch?v=E7p0vijKXn8.
21. https://www.forbes.com/sites/afdhelaziz/2019/01/10/the-power-of-purpose-how-x-prize-winners-skysourceskywater-alliance-turn-air-into-clean-water/.
22. https://www.waterfromair.co.za/water-from-air-aw3/.

23. https://pubs.acs.org/doi/abs/10.1021/acsami.7b17488.
24. Via interview with CNN: https://www.cnn.com/style/article/fog-harp-virginia-tech/index.html.
25. https://www.nytimes.com/2019/07/22/realestate/living-in-the-first-passive-house-high-rise.html.
26. https://www.nytimes.com/2019/07/22/realestate/living-in-the-first-passive-house-high-rise.html.
27. https://www.drawdown.org/solutions/coming-attractions/living-buildings.
28. See an interview with one of the building's designers here: https://www.phcppros.com/articles/8076-water-positive-buildings-are-bubbling-up.
29. https://otonomo.io/how-waycare-uses-otonomo-data-for-traffic-management-a-case-study/.
30. Some great artistic visuals of the plan here: https://inhabitat.com/first-smart-forest-city-in-mexico-will-be-100-food-and-energy-self-sufficient/.
31. https://www.traveler.es/experiencias/articulos/forest-city-en-china-proyecto-ciudad-al-norte-de-liuzhou/15339.
32. https://www.epa.gov/heat-islands/using-trees-and-vegetation-reduce-heat-islands.
33. https://www.nature.com/articles/s41467-019-10817-6.
34. https://news.utexas.edu/2019/05/29/a-rose-inspires-smart-way-to-collect-and-purify-water/.
35. https://www.bath.ac.uk/announcements/harnessing-the-sun-to-bring-fresh-water-to-remote-or-disaster-struck-communities/.
36. https://www.sciencenews.org/article/desalination-pours-more-toxic-brine-ocean-previously-thought.
37. https://pubs.acs.org/doi/10.1021/acs.estlett.9b00182.

CHAPTER 9: STEPPING BACK FROM THE BRINK: REVERSING CLIMATE CHANGE

1. https://www.nytimes.com/2018/11/19/science/climate-change-doom.html.
2. https://www.ipcc.ch/site/assets/uploads/sites/2/2019/02/SR15_Chapter2_Low_Res.pdf.
3. Minx, J. Luderer, G. Creutzig, F. et al. (2017). The fast-growing dependence on negative emissions Environ. Res. Lett. 12 035007.
4. For a rundown of scaling times, see: May, M. M. and Rehfeld, K.: ESD Ideas: Photoelectrochemical carbon removal as negative emission technology, Earth Syst. Dynam., 10, 1–7, https://doi.org/10.5194/esd-10-1-2019, 2019.
5. https://www.nap.edu/read/25259/chapter/10.
6. https://e360.yale.edu/digest/planting-1-2-trillion-trees-could-cancel-out-a-decade-of-co2-emissions-scientists-find.
7. Bastin, J. Finegold, Y. Garcia, C. et al. (2019). The global tree restoration potential. *Science*, 365 (6448), pp.76-79. DOI: 10.1126/science.aax0848.
8. https://onlinelibrary.wiley.com/doi/full/10.1111/gcb.14954.

9. https://science.sciencemag.org/content/366/6463/eaaz0388.
10. http://www.leeds.ac.uk/news/article/4555/tropical_forests_carbon_sink_already_rapidly_weakening.
11. https://www.biocarbonengineering.com/.
12. As their CEO told FastCompany: https://fastcompany.com/90329982/these-tree-planting-drones-are-firing-seed-missiles-to-restore-the-worlds-forests.
13. Via their website: www.Airseedtech.com.
14. https://www.dendra.io/blog/biocarbon-engineering-receives-us-2-5-million-in-investment-to-advance-drone.
15. Waite, C. van der Heijden, G. Field, R. et al. (2019). A view from above: Unmanned aerial vehicles (UAVs) provide a new tool for assessing liana infestation in tropical forest canopies. Journal of Applied Ecology. DOI: 10.1111/1365-2664.13318.
16. Gough, M. Atkins, J Fahey, R. et al. (2019). High rates of primary production in structurally complex forests. Ecology. DOI: 10.1002/ecy.2864.
17. https://www.wwf.de/fileadmin/user_upload/PDF/WWF-Globaler-Waldreport_BelowTheCanopy.pdf.
18. Fei, S. Morin, R. Oswalt, C. et al. (2019). Biomass losses resulting from insect and disease invasions in US forests. Proceedings of the National Academy of Sciences. DOI: 10.1073/pnas.1820601116.
19. Researchers via interview: https://www.agriland.ie/farming-news/native-species-have-a-higher-capacity-to-sequester-carbon/.
20. https://www.trilliontreecampaign.org/.
21. https://www.independent.co.uk/news/world/asia/china-tree-plant-soldiers-reassign-climate-change-global-warming-deforestation-a8208836.html.
22. Jones, M. Santin, C. van der Werf, G. et al. (2019). Global fire emissions buffered by the production of pyrogenic carbon. Nature Geoscience. DOI: 10.1038/s41561-019-0403-x.
23. https://www.independent.co.uk/news/world/asia/china-tree-plant-soldiers-reassign-climate-change-global-warming-deforestation-a8208836.html.
24. https://www.researchgate.net/publication/297301426_Mapping_opportunities_for_forest_landscape_restoration.
25. See several scientists' estimates in an article at The Guardian: https://www.theguardian.com/environment/2019/jul/04/planting-billions-trees-best-tackle-climate-crisis-scientists-canopy-emissions.
26. According to Deakin University marine ecologist Peter Macreadie, interview with ABC Australia: https://www.abc.net.au/news/science/2018-03-26/blue-carbon-mangroves-seagrass-fight-climate-change/9564096.
27. https://www.greenbiz.com/article/why-protecting-blue-carbon-storage-crucial-fighting-climate-change.
28. https://www.greenbiz.com/article/why-protecting-blue-carbon-storage-crucial-fighting-climate-change.
29. https://www.nature.com/articles/s41467-019-12176-8.
30. https://www.nature.com/articles/s41558-018-0090-4.
31. https://www.nature.com/articles/s41558-019-0485-x.

32. Froelich, H. Afflerbach, J. Frazier, M. et al. (2019). Blue Growth Potential to Mitigate Climate Change through Seaweed Offsetting. Current Biology. DOI: 10.1016/j.cub.2019.07.041.

33. http://www.ocean-based.com/technology/.

34. https://today.oregonstate.edu/news/smaller-expected-phytoplankton-may-mean-less-carbon-sequestered-sea-bottom.

35. https://www.drawdown.org/solutions/land-use/bamboo.

36. Bastin, J. Finegold, Y. Garcia, C. et al. (2019). The global tree restoration potential. *Science*, 365 (6448), pp.76-79. DOI: 10.1126/science.aax0848.

37. Dannenberg, M. Wise, E. Smith, W. (2019). Reduced tree growth in the semiarid United States due to asymmetric responses to intensifying precipitation extremes. Science Advances. DOI: 10.1126/sciadv.aaw0667.

38. Via interview: theguardian.com/environment/2019/apr/16/super-plants-climate-change-joanne-chory-carbon-dioxide.

39. For the science buffs: this is largely due to suberin's natural ability as a carbon polymer to resist short-term decomposition.

40. For a great review of what we know about this enzyme, I suggest: https://doi.org/10.1016/j.copbio.2017.07.017.

41. https://phys.org/news/2016-11-biologist-discusses-synthetic-metabolic-pathway.html.

42. Naseem, M. Osmanoglu, O. Dandekar, T. (2019). Synthetic Rewiring of Plant CO2 Sequestration Galvanizes Plant Biomass Production. Trends in Biotechnology. DOI: 10.1016/j.tibtech.2019.12.019.

43. M. Papanatsiou et al. (2019). Optogenetic manipulation of stomatal kinetics improves carbon assimilation, water use, and growth, Science. DOI: 10.1126/science.aaw0046.

44. May, M. Rehfeld, K. (2019). ESD Ideas: Photoelectrochemical carbon removal as negative emission technology. Earth Systems Dynamics 10, pp. 1-7. https://doi.org/10.5194/esd-10-1-2019.

45. Arborea's coverage in *The Guardian* was where I first noticed the technology (https://www.theguardian.com/environment/2019/apr/28/biosolar-leaf-project-targets-air-pollution-on-london-campus), but you could also head to their website for a FAQ: http://arborea.io/.

46. https://www.hypergiant.com/green/.

47. https://phytonix.com/.

48. In Nocera's research, via interview: https://www.forbes.com/sites/samlemonick/2017/04/05/bionic-leaf-makes-fertilizer-from-sunlight-and-air/.

49. See the interview with May in *Popular Science*: https://www.popsci.com/artificial-trees-photosynthesis-climate-change-carbon.

50. As described in a PR post from the university: https://today.uic.edu/moving-artificial-leaves-out-of-the-lab-and-into-the-air.

51. https://pubs.acs.org/doi/10.1021/acssuschemeng.8b04969.

52. https://sequestration.mit.edu/tools/projects/.

53. https://www.theguardian.com/environment/2018/oct/17/carbon-capture-technology-climate-change-solutions.

54. https://carbonengineering.com/frequently-asked-questions/.

55. Valuations are also from Carbon Engineering's FAQ.

56. https://www.climeworks.com/climeworks-response-to-the-ipcc-report/.

57. https://www.nytimes.com/2019/02/12/magazine/climeworks-business-climate-change.html.

58. https://apnews.com/Business%20Wire/f0edc67131d04d2dbd7a531ebdecd422.

59. Ibid.

60. https://www.technologyreview.com/s/612928/one-mans-two-decade-quest-to-suck-greenhouse-gas-out-of-the-sky/.

61. https://web.stanford.edu/group/efmh/jacobson/Articles/I/AirCaptureVsWWS.pdf.

62. Researcher via interview: https://www.greenbiz.com/article/case-investing-direct-air-capture-just-got-clearer.

63. https://news.engineering.utoronto.ca/out-of-thin-air-new-electrochemical-process-shortens-the-path-to-capturing-and-recycling-co2/.

64. Valencia, L. Rosas, W. Aguilar-Sanchez, A. et. al. (2019). Bio-based Micro-/Meso-/Macroporous Hybrid Foams with Ultrahigh Zeolite Loadings for Selective Capture of Carbon Dioxide. ACS Appl. Mater. Interfaces, 11 (43). https://doi.org/10.1021/acsami.9b11399.

65. http://dx.doi.org/10.1038/s41893-019-0299-x.

66. Voskian, S. Hatton, T. (2019). Faradaic electro-swing reactive adsorption for CO_2 capture. Energy & Environmental Science. DOI: 10.1039/C9EE02412C.

67. https://vimeo.com/368583616.

68. Hawkins' transcript lines are located in an iNews piece, but consider looking up his contributions in other conferences: https://inews.co.uk/news/environment/climate-change-carbon-capture-technology-634074.

69. Hepburn, C. Adlen, E. Beddington, J. et al. (2019). The technological and economic prospects for CO_2 utilization and removal. *Nature*, 575 (7781), p.87 DOI: 10.1038/s41586-019-1681-6.

70. https://aircompany.com/.

71. https://www.10xbeta.com/footprintless/.

72. https://www.cleano2.ca/products/carbinx-c1-version-3-3.

73. https://www.iea.org/newsroom/news/2018/april/cement-technology-roadmap-plots-path-to-cutting-co2-emissions-24-by-2050.html.

74. See New Jersey's Solidia, for estimates: https://www.solidiatech.com/.

75. https://www.solidiatech.com/solutions.html.

76. https://www.co2concrete.com/carbon-capture-process/.

77. https://doi.org/10.1038/s41586-019-1681-6.

78. Sathre, R. & O'Connor, J. Meta-analysis of greenhouse gas displacement factors of wood product substitution. Environ. Sci. Policy 13, 104–114 (2010).

79. https://www.nytimes.com/2019/11/20/style/engineered-wood-tower-construction.html.

80. https://www.nytimes.com/2019/11/20/style/engineered-wood-tower-construction.html.
81. https://www.nature.com/articles/s41893-019-0462-4.
82. https://www.iis.u-tokyo.ac.jp/en/news/3241/.
83. https://dx.doi.org/doi:10.1021/acs.accounts.9b00405.
84. https://www.ipcc.ch/report/srccl/.
85. https://linkinghub.elsevier.com/retrieve/pii/S0959378016303399.
86. https://pubs.rsc.org/en/content/articlelanding/2018/ee/c7ee02342a.
87. https://iopscience.iop.org/article/10.1088/1748-9326/11/11/115007.
88. https://pubs.rsc.org/en/content/articlehtml/2018/ee/c7ee02342a.
89. https://www.carbonbrief.org/worlds-soils-have-lost-133bn-tonnes-of-carbon-since-the-dawn-of-agriculture.
90. See also a good reference from the Earth Institute, Columbia University: https://blogs.ei.columbia.edu/2018/02/21/can-soil-help-combat-climate-change/.
91. https://www.nature.com/articles/s41893-020-0491-z.
92. The UN's Environment Programme estimates that peatlands store twice as much carbon as all the world's forests.https://www.unenvironment.org/news-and-stories/story/peatlands-store-twice-much-carbon-all-worlds-forests.
93. https://www.drawdown.org/solutions/food/silvopasture.
94. https://royalsocietypublishing.org/doi/abs/10.1098/rstb.2007.2184.
95. https://www.commondreams.org/views/2019/08/12/regenerative-agriculture-key-sustainable-climate-and-food-system.
96. See: https://www.ucdavis.edu/news/compost-key-sequestering-carbon-soil.
97. Nottingham, A. Whitaker, J. Ostle, N. et al. (2019). Microbial responses to warming enhance soil carbon loss following translocation across a tropical forest elevation gradient. Ecology Letters. DOI: 10.1111/ele.13379.
98. https://www.ncbi.nlm.nih.gov/pubmed/28421095.
99. http://www.fao.org/fileadmin/templates/agphome/documents/scpi/PerennialPolicyBrief.pdf.
100. https://e360.yale.edu/features/perennial_rice_in_search_of_a_greener_hardier_staple_crop.
101. https://e360.yale.edu/features/with-new-perennial-grain-a-step-forward-for-eco-friendly-agriculture.
102. https://www.drawdown.org/solutions/food/regenerative-agriculture.
103. For a fantastic summation of this concept, see this piece from Russell McLendon at MNN: https://www.mnn.com/earth-matters/climate-weather/blogs/the-pliocene-called-it-wants-its-co2-levels-back.
104. More exposition here: https://www.theguardian.com/science/2019/apr/03/south-pole-tree-fossils-indicate-impact-of-climate-change.
105. https://www.pnas.org/content/114/44/11645.
106. https://www.nature.com/articles/s41467-019-10842-5.
107. https://link.springer.com/article/10.1007/s40641-018-0104-3.

CHAPTER 10: DAMAGE CONTROL

1. Contained in: Minteer, B. Pyne, S. (eds.) (2015). After Preservation: Saving American Nature in the Age of Humans. University of Chicago Press.
2. http://www3.weforum.org/docs/WEF_The_New_Plastics_Economy.pdf.
3. https://www.businessinsider.com/plastic-in-ocean-outweighs-fish-evidence-report-2017-1.
4. https://wwf.panda.org/wwf_news/press_releases/?348337/Revealed-plastic-ingestion-by-people-could-be-equating-to-a-credit-card-a-week.
5. https://pubs.acs.org/doi/full/10.1021/acs.est.9b01517.
6. https://ipbes.net/news/Media-Release-Global-Assessment.
7. According to Dr. Jennifer Provencher of the Canadian Wildlife Service, cited by The Independent: https://www.independent.co.uk/environment/plastic-bird-eggs-pollution-arctic-ocean-chemicals-phthalates-research-a8783061.html.
8. Gerona, R. Saal, F. Hunt, P. (2019). BPA: have flawed analytical techniques compromised risk assessments? The Lancet Diabetes & Endocrinology. DOI: 10.1016/S2213-8587(19)30381-X.
9. https://pubs.acs.org/doi/10.1021/acs.est.7b02368.
10. https://theoceancleanup.com/.
11. https://www.kickstarter.com/projects/367174604/floating-coconet-cleaning-plastic-rivers.
12. https://www.cnn.com/2019/08/01/us/irish-teen-wins-google-science-fair-trnd/index.html.
13. https://thegreatbubblebarrier.com/en/.
14. https://news.wsu.edu/2020/03/13/invisible-plastics-water/.
15. https://www.cnn.com/2019/10/02/tech/ocean-cleanup-catching-plastic-scn-trnd/index.html.
16. https://www.cnn.com/2018/10/30/middleeast/wasteshark-drone-dubai-marina/index.html.
17. https://www.theguardian.com/us-news/2019/dec/31/ocean-plastic-we-cant-see.
18. https://www.nature.com/articles/s41598-019-44117-2.
19. https://www.lunduniversity.lu.se/article/brain-damage-in-fish-affected-by-plastic-nanoparticles.
20. Links to the pertinent research and more on the campaign's landing: https://gojelly.eu/.
21. http://news.mit.edu/2019/mit-robots-can-sort-recycling-0416.
22. https://www.nytimes.com/2019/05/23/world/asia/philippines-canada-trash.html.
23. https://www.sciencedirect.com/science/article/pii/S0304389419305060.
24. https://carbios.fr/en/technology/biodegradation/.
25. https://carbios.fr/en/technology/biorecycling/.
26. https://www.fastcompany.com/90417038/in-this-biorecycling-factory-enzymes-perfectly-break-down-plastic-so-it-can-be-used-again.
27. https://www.nature.com/articles/s41586-020-2149-4.

28. https://www.kiverdi.com/reverse-plastics.
29. Their estimates provided to FastCompany: https://www.fastcompany.com/90420730/los-angeles-is-testing-plastic-asphalt-that-makes-it-possible-to-recycle-roads.
30. https://pubs.acs.org/doi/10.1021/acssuschemeng.8b03841.
31. Zhang, Y. Duan, D. Lei, H. et al. (2019). Jet fuel production from waste plastics via catalytic pyrolysis with activated carbons. Applied Energy, 251. DOI: 10.1016/j.apenergy.2019.113337.
32. See their commentary on the university's PR post: https://news.wsu.edu/2019/06/04/plastic-water-bottles-may-one-day-fly-people-cross-country/.
33. Gazi, S. Dokic, M. Chin, K. (2019). Visible Light–Driven Cascade Carbon–Carbon Bond Scission for Organic Transformations and Plastics Recycling. Advanced Science. DOI: 10.1002/advs.201902020.
34. https://apeelsciences.com/science/.
35. https://twentyfiftyfork.com/.
36. Outlined by British Airways after selecting this research for their cleaner fuel awards: https://mediacentre.britishairways.com/pressrelease/details/86/2019-319/11024.
37. https://www.cnbc.com/2020/01/06/jetblue-will-be-carbon-neutral-on-all-domestic-flights-by-july-2020.html.
38. https://www.neste.us/about-neste/sustainability/cleaner-solutions.
39. https://www.uop.com/processing-solutions/renewables/green-jet-fuel/.
40. A figure I first found on Wired, backed up by the linked report. https://www.wired.co.uk/article/aviation-biofuels-carbon-emissions-waste-algae.
    The report: International Air Transport Association (IATA). https://www.iata.org/pressroom/facts_figures/fact_sheets/Documents/fact-sheet-climate-change.pdf.
41. https://www.nature.com/articles/s41893-018-0172-3.
42. Brinson, B. Kumar, A. Hauge, R. (2019). From Newspaper Substrate to Nanotubes—Analysis of Carbonized Soot Grown on Kaolin Sized Newsprint. C — Journal of Carbon Research, 5 (4), p.66 DOI: 10.3390/c5040066,
43. https://widgets.weforum.org/techpioneers-2019/companies/black-bear-carbon/.
44. Cited in the company's Tech Pioneers landing on WEF, link above.
45. See an insightful analysis on Greenbiz here, including their phrasing: https://www.greenbiz.com/article/climate-solutions-depend-rare-earths-heres-how-they-can-be-sourced-responsibly.
46. http://dx.doi.org/10.1126/science.1159196.
47. https://unesdoc.unesco.org/ark:/48223/pf0000265625.
48. In conversation with Nexus Media: https://nexusmedianews.com/environmental-scientists-want-help-coping-with-their-grief/.
49. Price, N. Muko, S. Legendre, L. et al. (2019). Global biogeography of coral recruitment: tropical decline and subtropical increase. Marine Ecology Progress Series, 621 (1). DOI: 10.3354/meps12980.
50. https://www.npr.org/2019/10/16/766200948/trees-that-survived-california-drought-may-hold-clue-to-climate-resilience.

51. https://www.nap.edu/catalog/25279/a-research-review-of-interventions-to-increase-the-persistence-and-resilience-of-coral-reefs.
52. Martinez, A. Crook, E. Barshis, D. et al. (2019). Species-specific calcification response of Caribbean corals after 2-year transplantation to a low aragonite saturation submarine spring. Proceedings of the Royal Society B: Biological Sciences, 2019; 286 (1905): 20190572 DOI: 10.1098/rspb.2019.0572.
53. Camp, E. Edmondson, J. Doheny, A. (2019). Mangrove lagoons of the Great Barrier Reef support coral populations persisting under extreme environmental conditions. Marine Ecology Progress Series. DOI: 10.3354/meps13073.
54. https://www.nap.edu/read/25279/chapter/2#3.
55. Read up on the development here: https://phys.org/news/2019-11-nurseries-robotic-fleet-coral-babies/.
56. https://dlnr.hawaii.gov/ais/invasivealgae/supersucker/.
57. Cited in Christopher Preston's 2019 The Synthetic Age.
58. https://news.utexas.edu/2020/01/30/bacteria-engineered-to-protect-bees-from-pests-and-pathogens/.
59. https://jamanetwork.com/journals/jama/article-abstract/2747669.
60. https://pubs.acs.org/doi/10.1021/acs.est.8b04811.
61. https://apnews.com/77d1e355d6714bb0b47f66d560f119f9.
62. https://chakr.in/.
63. https://pubs.acs.org/doi/10.1021/acs.est.7b00891.
64. https://report.ipcc.ch/sr15/pdf/sr15_chapter4.pdf.
65. http://iopscience.iop.org/article/10.1088/1748-9326/aae98d/meta.
66. https://www.nature.com/news/iron-dumping-ocean-experiment-sparks-controversy-1.22031.

CHAPTER 11: POSSIBLE FUTURES

1. https://cuesa.org/learn/how-far-does-your-food-travel-get-your-plate.
2. https://www.npr.org/2018/12/06/674075032/california-gives-final-ok-to-requiring-solar-panels-on-new-houses.
3. https://news.rice.edu/2019/10/14/synthetic-cells-make-long-distance-calls/.
4. https://www.c40.org/press_releases/30-of-the-world-s-largest-most-influential-cities-have-peaked-greenhouse-gas-emissions.
5. https://www.nature.com/articles/s41558-019-0419-7.
6. https://www.un.org/development/desa/en/news/population/2018-revision-of-world-urbanization-prospects.html.
7. Huang, K. Li, X. Liu, X. et al. (2019). Projecting global urban land expansion and heat island intensification through 2050. Environmental Research Letters, 14 (11): 114037 DOI: 10.1088/1748-9326/ab4b71.
8. https://www.iea.org/articles/global-co2-emissions-in-2019.
9. https://www.irena.org/publications/2019/Apr/Global-energy-transformation-A-roadmap-to-2050-2019Edition.
10. https://www.nature.com/articles/s41558-020-0740-1.

11. https://www.rethinkx.com/food-and-agriculture.

12. https://www.theguardian.com/science/2010/may/20/craig-venter-synthetic-life-form.

13. https://www.acs.org/content/acs/en/pressroom/presspacs/2020/acs-presspac-april-8-2020/harnessing-the-power-of-electricity-producing-bacteria-for-programmable-biohybrids.html.

14. https://www.nature.com/articles/s41467-019-12190-w.

15. https://warwick.ac.uk/newsandevents/pressreleases/an_artificial_intelligence.

16. https://www.pnas.org/content/early/2020/01/07/1910837117.

17. https://www.energy.gov/eere/solar/sunshot-2030.

18. See a good write-up on this concept here, including suggestions for policymakers: https://www.wri.org/blog/2020/01/5-ways-us-government-can-kickstart-carbonshot-remove-carbon-atmosphere.

19. https://cires.colorado.edu/news/international-ozone-treaty-stops-changes-southern-hemisphere-winds.

20. https://iopscience.iop.org/article/10.1088/1748-9326/ab4874.

21. https://www.nature.com/articles/nclimate2921.

22. If looked up via Twitter, you can also see a handy infographic: https://twitter.com/KHayhoe/status/1216491161566707712.

23. https://www.theguardian.com/environment/2015/jul/10/denmark-wind-windfarm-power-exceed-electricity-demand.

24. https://www.bloomberg.com/news/articles/2019-01-02/world-s-biggest-ultra-high-voltage-line-powers-up-across-china.

25. DOI:10.1016/j.enconman.2016.01.019. https://inis.iaea.org/search/search.aspx?orig_q=RN:48003147.

26. See the discussions from project engineers in mainstream media here: https://www.forbes.com/sites/quora/2016/09/22/we-could-power-the-entire-world-by-harnessing-solar-energy-from-1-of-the-sahara/.

    And here: https://theconversation.com/should-we-turn-the-sahara-desert-into-a-huge-solar-farm-114450.

27. https://link.springer.com/article/10.1007/s11027-019-9847-y.

28. Ibid—see study's notes.

29. Grainger, A. Iverson, L. Marland, G. et al. (2019). Comment on "The global tree restoration potential". *Science*, 366 (6463). DOI: 10.1126/science.aay8334.

30. Possner, A. Caldeira, K. (2017). Geophysical potential for wind energy over the open oceans. Proceedings of the National Academy of Sciences. DOI: 10.1073/pnas.1705710114.

INDEX

ABOUT THE AUTHOR

MARC SCHAUS is a professional research specialist across the sciences for research ventures, craft product manuals, and policymakers. He is the author of *Post Secular: Science, Humanism and the Future of Faith,* and has written articles that have appeared in *Areo Magazine, Free Inquiry Magazine,* The Huffington Post, Patheos, and the academic journal *Antennae.* He currently resides in Ontario, Canada.